The DARWIN Papers

This collection was compiled in response to the difficulty people have been experiencing in finding my papers on Darwin. The commercial system for academic journals is certainly partly to blame, and open access publishing is still in its infancy. Even so, not everyone has access to, or knowledge of the online literature resources where most of these papers have previously appeared. That is why they are now collected for your convenience. However, please note that in order to satisfy copyright requirements, the versions here are earlier drafts, mostly written in 2009, and differ sufficiently from their published descendants. Also added are a handful of my later, relevant, blog posts written between 2010 and 2012, plus some excellent artwork. Admittedly, it doesn't make for a very cohesive collection; the range of topics is too far reaching. However, Darwin is at the core of each paper, and now at least they can be found in one place.

At the time of writing, my next book on Darwin is *The Dissent of Man*, currently being funded via online crowd-funding by Unbound publishers. Please take a look and help out if you can. Thank you.

http://unbound.co.uk/books/the-dissent-of-man/

JF Derry
Edinburgh 2013

First published in 2013

ISBN: 978-1-291-32590-4

All rights reserved

© JF Derry, 2013

JF Derry is hereby identified as the author of this work in accordance with Section 77 of the Copyright, Designs and Patents Act, 1998

Except:
Water location, piospheres and a review of evolution in African ruminants co-written with Andy Dougill

All original artwork ©Simon Gurr, 2009, 2012 http://gurr.eu

Photograph copyrights as accredited on individual images.

All rights reserved. No part of this publication may be reproduced, stored in a retrieval system, or transmitted, in any form or by any means, electronic, mechanical, recording or otherwise without prior permission of the publishers.

---

by the same author

Piospheres

Darwin in Scotland

Serial Killers

Loch Ness Monster

The Dissent of Man
http://unbound.co.uk/

The DARWIN Papers

DARWIN & THE UNIVERSITY OF EDINBURGH	P5
DARWIN / MONKEY	P12
THINKING KONG	P19
BRAVO EMMA!	P24
RICH PICKINGS	P39
DARWIN IN DISGUISE	P51
WATER LOCATION AND EVOLUTION	P69
PENIS SIZE AND PEACOCKS' EYES	P107
WARS OF THE WORD	P112
DOES GENOMICS NEED DARWIN?	P124
ONE GIANT LEAP FOR MANKIND	P134
A ROAD JOURNEY	P154
DISSECTING THE DISSENT OF MAN	P162

JF Derry

Darwin and the University of Edinburgh

Edinburgh was vital preparation for Darwin's voyage aboard the Beagle.

Darwin's history at the University could have been made more of, given the bicentenary celebrations, and that his two-year sojourn here shaped his view of the world, and vitally, an ability to measure it: the University of Edinburgh seems to have been a bountiful source of influences on the nature loving but unambitious sixteen year old.

1.Darwin's Edinburgh

Sent to study medicine in 1825, Darwin's first year was most notable for his prodigious book borrowing while sharing digs with his brother Erasmus "Eras" Alvey, who, as part of his own medical degree, was embarking on a year of hospital residency of which he would only complete four months. When Eras left, Darwin was abandoned to suffer the "intolerably dull" lectures, "with the exception of those on chemistry by [Professor] Hope", whereas "Dr. Duncan's lectures on Materia Medica at 8 o'clock on a winter's morning are something fearful to remember. Dr. Munro made his lectures on human anatomy as dull, as he was himself, and the subject disgusted me". More disgusting was having to attend two unanesthetized operations, something he never repeated, nor needed to: "The two cases fairly haunted me for many a long year".

To clear his head, he took an energetic walking holiday in North Wales that summer, where he became further inspired about wildlife before returning to a more sociable

second year in Edinburgh. A freed black slave John Edmonstone taught him taxidermy, invaluable on the Beagle voyage and Leonard Horner "took me once to a meeting of the Royal Society of Edinburgh, where I saw Sir Walter Scott in the chair as President".

Notably, he joined Professor Robert Jameson's Plinian Society and presented a paper there on the marine biology of the Firth of Forth as a result of collecting in tidal pools near Prestonpans with possibly the single most important influence, Dr Robert Edmund Grant, about whom he recalled, "I knew him well; he [...] burst forth in high admiration of Lamarck and his views on evolution [...] it is probable that the hearing rather early in life such views maintained and praised may have favoured my upholding them under a different form in my *Origin of Species*".

2. Athens of the North

Edinburgh earned the name "Athens of the North" for more than the Parthenon replica on Carlton Hill; it played a central role in the Scottish Enlightenment, with many of the movement's key figures being either in attendance at the University of Edinburgh or living in the city. In this latter group was David Hume, exponent of modern scientific empiricism, upon which all scientific truth has rested ever since. The Pyrrhonian skepticism adapted by Hume thus defines a clear line from Ancient Greece, via the Scottish Enlightenment, straight to the Victorian scientific naturalism that Darwin was first exposed to in Edinburgh.

3.All Greek

Evolution also began with the Greeks, as did natural selection, but Darwin is the true "Father of Evolution" because of the *way* in which he formed his ideas: using his own acute powers of observation in tandem with the stringent scientific method derived from antiquity. Darwin's discoveries were so powerfully robust that they continue to be applied across a broad range of sciences and are fundamental to every biological discipline. The popularity of his ideas was no doubt initially assisted by the air of discovery in the Victorian era, paved by the Age of Enlightenment, but the persistent longevity of his ideas is beyond doubt. No one can claim that they haven't survived the test of time.

4.Hammer Time

Darwin's gradualism came from his appreciation of geological time, and again, Edinburgh was the source. City-born James Hutton was fascinated by local rock formations but just couldn't connect what he was seeing with the widely-accepted Neptunism, the belief that rocks crystallised out of Earth's early oceans. Instead he proposed a Plutonist argument that rock was manipulated and mixed and eroded, over a great period by volcanism. Suddenly the Young Earth, was insufficient to accommodate the evidence and the vital concept of "deep time" was born.

Darwin absorbed Hutton's influence while his own formal geologic training did little to instruct him. Attending Robert Jameson's courses during his second year in Edinburgh he should have been stimulated by lessons on stratigraphic geology, largely instigated by Georges Cuvier. Instead, he was bored by Jameson's course, although it did perhaps seed in Darwin a title for *The Origin...*: Jameson's 1826 course in zoology concluded with lectures on the philosophy of zoology, starting with *Origin of the Species of Animals*. In addition, the immediate benefit of attending was to give him a chance to help with the collections in the University museum, a flashback to his indiscriminate collecting as a childhood hobby.

It's little known that as an assistant at the University Museum Darwin had use of a desk. The desk he was allocated sits at the bottom of a short stairwell in a building at the university's Old College, which since 1975 has housed the Talbot Rice Gallery. Now, that staircase is special. It is an elegant, sweeping, spiral one. So, leaning back, hands behind his head, gazing absentmindedly cupola-wards, Darwin's gaze would have followed the parallel lines of a double helix towering above him, but how fundamentally relevant it would be, he dnae ken. What sweet, delicious, wrought irony.

Darwin / Monkey

Edina! Scotia's darling seat! ... There Learning, with his eagle eyes, Seeks Science in her coy abode.
from Address To Edinburgh by Robert Burns (1759–1796)

There is great excitement at the bicentenary of Charles Darwin's birth and the 150-year anniversary of *The Origin of Species*, both in 2009. Understandably, there has been much celebrating Robert Burns' 250th birthday this year, yet, to date, little has been done to reaffirm Scotland's, and particularly Edinburgh's role in Darwin's history.

Also, the celebrations are late; 1856 was a momentous year towards the publication of Darwin's ideas. In that year he shared his ideas with colleagues, and the response was largely positive giving Darwin an important confidence boost. Having found acceptance of his ideas, the nudge he needed to publish them arrived a short time later when Charles Lyell, a Scot, alerted him to similar findings in the work of Alfred Russel Wallace, himself born in Welsh Monmouthshire, but to an English mother and a father claiming direct descent from William Wallace. Thus, they were Scottish sources that inspired publication of *The Origin of Species...*, and it was even a Scottish source from which that famous book's very title likely was seeded: his 1826 course in zoology at Edinburgh University concluded with lectures on the philosophy of zoology, starting with *Origin of the Species of Animals*.

Most important for Darwin's skills as a naturalist, it was time spent in Edinburgh and the Highlands of Scotland that informed and trained the young man in preparation for his legendary voyage aboard the *Beagle* and his subsequent career in science. Darwin would not have developed his theories if he had not attended Edinburgh University. His formal tuition there didn't amount to much, but through interaction with his tutors, peers and extracurricular groups, Darwin was exposed to an ethos of naturalistic philosophy rooted in the Scottish Enlightenment, and by direct descent, the Ancient Greeks.

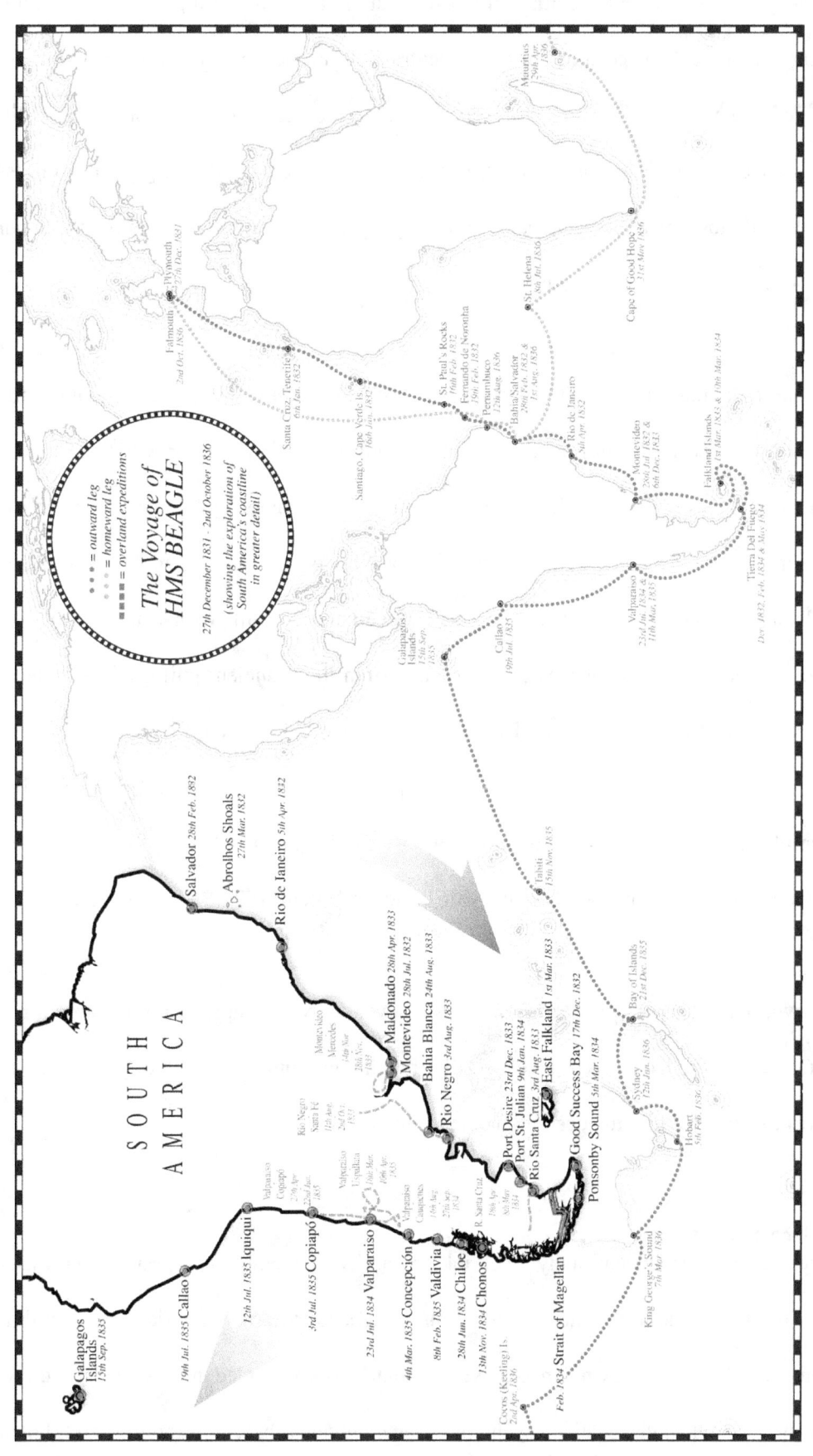

Athens of the North
Edinburgh wasn't the first Scottish university but its academic pedigree took the fore for having a cosmopolitan perspective and a central role in the Scottish Enlightenment. While this movement riveted in moral philosophy, history and economics, and surviving until the latter half of the 18th century, can be traced to the University of Glasgow as well as the Aberdeen Philosophical Society, many of the key figures were at the University of Edinburgh, or outside academia, but living in the city.

Outstanding in this latter group was the giant David Hume (1711-1776), exponent of modern scientific empiricism (founded in "experience and observation"), upon which all scientific truth has rested ever since. In Hume's time, Edinburgh was famously known as the "Athens of the North". That Athenian parallel with the classical period extends to the influence that Scotland had on the modern world, just as Greece had once enlightened the ancient world. Other associations forge lineages of inheritance from those ancient philosophies. Hume, for one, was directly influenced by Pyrrho (ca. 365–ca. 275 BCE) founder of Pyrrhonian skepticism (in contrast to Plato's Academic skepticism), and of course also indirectly, by the accumulation of independent, rationalism across the intervening two millennia, a line that started where philosophy itself begins, with Thales of Miletus (624 BC–ca. 546 BC).

If Darwin *had* bypassed Scotland, going, as he eventually did, straight up to Cambridge, his influences would have been theologically based, and unlikely to have given him the perspective that led him to question the prevailing doctrine.

Too free or not too free?
Darwin came from an unusually freethinking family of medics. His grandfather was in a minority as an evolutionist and the whole family tended towards Unitarianism. The ultimate impact on Darwin of having mixed with similarly open-minded naturalists at Edinburgh cultivated in him a interrogative psyche that was not evident before his coming to Edinburgh, and one that was to remain with him for the rest of his life.

Darwin's grandfather Erasmus, father Robert, uncle also Charles and brother Erasmus, all attended the Medical School in Edinburgh, benefiting from the parallel teaching of both theory and practise that was not afforded by anywhere else at the time, and so making it the best medical education from the mid-18th century onwards. Robert alone went on to practise medicine, although Erasmus certainly treated family members. However, medicine was not for Darwin, although an education balanced by theory coupled with Hume's empiricism resonates throughout his future life's work.

Son Francis wrote about Darwin's approach to work, a couple of years after his passing, "He often said that no one could be a good observer unless they were an active theoriser [...] it was as tho' he were charged with theorising power ready to flow into any channel [...] but fortunately his richness of imagination was equalled by his power of judging & condemning the thoughts that occurred to him [...] and so it happened that he was willing to test what would seem to most people not at all worth testing. These rather wild trials he called "fool's experiments" & enjoyed extremely [...] his wish to test the most improbable ideas. This wish was very strong in him and I can remember the way he said "I shan't be easy till I've tried it" — as if an outside force were compelling him".

Willingness to test. Without this state of mind, Darwin may well have never questioned the current understanding, and without his explanation of evolutionary processes, he would never have reached the uncomfortable but inevitable contradiction of the widely accepted creationist account of human origins. Darwin's conclusions were especially uncomfortable on a personal level. They were a direct threat to his wife's piousness which produced in him a terrible conflict between his professional intent, and the protection of her feelings.

Wilde life
Art is born of the observation and investigation of nature.
Cicero (106 BC - 43 BC)

Great art picks up where nature ends.
Marc Chagall (1887 – 1985)

If Oscar Wilde wasn't far off the mark when he wrote, "Life imitates art far more than art imitates Life", then we have, within a small exedra in the Ashworth Labs at the University, a piece of art, a famous statuette (pictured), that beautifully captures Darwin's well documented struggle. It is the "Affe mit Schädel" ("Ape with Skull") by the German sculptor Hugo Rheinhold. A duplicate takes pride of place at the Aberdeen Medico-Chirurgical Society.

Photo: Steven Hay

It is of a Common Chimpanzee (*Pan troglodytes*) sitting atop a higgledy-piggledy pile of books and manuscripts. The subject is likely female, based on her smaller brow ridge, and in the absence of any contradictory, genital evidence. She has in her right forepaw a human skull, echoing the scene in Shakespeare's "Hamlet" where the Prince of Denmark bereaves Yorick ("Alas, poor Yorick! I knew him..."). She cradles her chin with her other forepaw in a contemplative posture. Her left foot holds the shin of her right leg as if to steady it or support the callipers held by her right foot. A spine of one of the closed books reads "DARWIN". The book open at her feet and facing the viewer has a single inscription on the right hand page, "ERITIS SICUT DEUS". This is a quote from Genesis 3.5, when the serpent is enticing Eve to eat of the Tree of Knowledge, promising, "And ye shall be as God [knowing good and evil]"). However, the second half of this quote is missing, ripped from the lower half of the page.

Thinking Kong
What inspired Rheinhold in making his sculpture is unknown. It has obvious parallels with Auguste Rodin's "The Thinker", but it is perhaps surprising to discover that while Rodin had developed his statue as early as 1880, it was not cast into bronze and displayed until after the "Affe mit Schädel" had debuted in 1893 at the Great Berlin Art Exhibition.

A large part of the statuette's popularity, are the myriad of possible interpretations. A message explicitly made elsewhere by Rheinhold cautions against imprudent use of technology. In the case of the "Affe mit Schädel", the excised Biblical quote suggests that good and evil cannot be known, or distinguished. With the ape's study, the library of books and the calliper instruments, the suggestion is that the statuette is warning against the application of rationalism in the absence of morality.

The "Affe mit Schädel" also deals with mortality. Specifically, when a human is depicted holding a skull it is usually about the inevitability of death. But, it is something quite different for our ape who is engaged in assessment and measurement. The countenance is not one of sorrow nor melancholy, but studious indifference, or even whimsy.

The Darwin Monkey
Upon its debut, the "Affe mit Schädel" became noticed by the Gladenbeck foundry who purchased the rights and featured it as a bronze in their catalogue. Although it became popular through its quirky originality, the full-size moulds disappeared when the foundry closed in 1926. The statuette in part symbolizes Darwin's own studious career: his engagement with the question of human origins, his empiricism, the exacting inspection of his own ideas informed through his assessment of other literature, and ultimately his own realisation of the conflict between evolution and creationism. A scholarly, non-human hominid acts as a powerful reinforcement of our primitive ancestry. In this sense, the "Affe mit Schädel" is not so much Darwinian, as *it is* Darwin.

Thinking Kong

Art is born of the observation and investigation of nature.
Cicero (106 BC - 43 BC)

Great art picks up where nature ends.
Marc Chagall (1887 – 1985)

The artist Alistair Gentry once discussed Darwin's relationship with the artistic elements of society, "Darwin himself was also following tropes and imagery that already present in the culture, not only other scientists or proto-scientists but also fiction and philosophical texts from the mid-1800s onwards that were obsessed by the idea of degeneration or devolution [...] *e.g.* Poe's *Murders in the Rue Morgue* in which the crimes are committed by an ape human enough to lust and mutilate, but not human enough to control its savagery, or Shelley's monster, a refined and sensitive intellect destined to commit murderous and loathsome acts [...] I'd consider the influence of art and fiction on Darwin in addition to vice versa. In fact I think there's more traffic in the former direction."

Similarly, in his "The Decay Of Lying - An Observation", Oscar Wilde also claimed that, "Life imitates art far more than art imitates Life".

Accepting that Science also feeds the Arts, the suggestion is that Art may act in a bidirectional communion with Science, better allowing us a balanced perspective: does Art bring values to scientific objectivity? The possibility is certainly suggested by an extraordinary piece of art that we have at the University. It beautifully imagines Darwin's insight within *The Descent of Man*, then extends the metaphor to consider the ethics of scientific progress. Meanwhile it's iconography is immediately redolent of Darwin's personal struggle over how his findings challenged orthodoxy.

Within a small exedra (a semicircular recess crowned by a half-dome), in the main stairwell of the Ashworth Labs (University's King's Buildings), there is a statuette of a Common Chimpanzee (*Pan troglodytes*) sitting atop a higgledy-piggledy pile of books and manuscripts. The subject appears female, and she has in her right forepaw a human skull,  echoing the scene in Shakespeare's "Hamlet" where the Prince of Denmark bereaves Yorick ("Alas, poor Yorick! I knew him..."). She cradles her chin with her other forepaw in a contemplative posture. Her left foot holds the shin of her right leg, steadying or supporting the callipers held by her right foot. A spine of one of the closed books reads "DARWIN". The book open at her feet and facing the viewer has a single inscription on the right hand page, "eritis sicut deus". This is a Biblical quote from Genesis 3.5, when the serpent is enticing Eve to eat of the Tree of Knowledge of Good and Evil in the Garden of Eden against the Lord's command. The serpent is promising, "And ye shall be as God *knowing good and evil*"), however the second half of this quote is missing, ripped from the lower half of the page.

This famous statuette is the "Affe mit Schädel" ("Ape with Skull") by the late-19th century German sculptor Hugo Rheinhold, and was first exhibited at the 1893 Great Berlin Art Exhibition.

But what does it mean?

One of the great things about the statuette, and a large part of its attraction, are the myriad of possible interpretations. An explicit message from another of Rheinhold's works ("Dynamite in the Service of Mankind") cautions against imprudent use of technology. Likewise, in the case of the "Affe mit Schädel", the excised quote suggests that good and evil cannot be known, or told apart. So, along with the ape's study, the library of books and the calliper instruments, the suggestion is that the statuette is warning against the application of rationalism in the absence of morality. This is a concern that has often been directed at Science, for example, nuclear technology, genetic engineering, and the recent proposed banning of primate experimentation.

Alternatively, the statuette in part symbolizes Darwin's own studious career: his engagement with the question of human origins, his empiricism, the exacting inspection of his own ideas informed through his assessment of other literature, and ultimately his own realization of the conflict between evolution and creationism.

The statuette turns the tables, interchanging a non-human hominid in what we know is a position only possible for humans. It acts as a powerful reinforcement of our primitive ancestry, a fact expertly confirmed by Charles Darwin.

In this sense, the "Affe mit Schädel" is a clever integration of Art and Science, but it is not a Darwinian piece of Art, as much as *it is* Darwin.

Bravo Darwin! Emma's Pianos And Music In The Home And Work Of Charles Darwin

Music washes away from the soul the dust of everyday life.
Berthold Auerbach (1812–1882)
German-Jewish poet and author

Introduction

The profusion of material being produced in this anniversary year leaves little new to be said about Charles Darwin. So, instead I'm going to focus on an unusual selection of Darwinian objects that does not receive much attention, but one that I hope will strike a chord with you: Emma Darwin's pianos. This article is offered as a musical note on Darwin, having a structure intended to reflect, but might not quite conform to, the sonata-allegro form that would have been familiar to Emma. Traditionally, sonatas are compositions in three sections: the exposition, the theme; development, one or more themes, often in juxtaposing keys to the dominant thematic key; and the recapitulation, an altered repeat of the exposition.

Painting of Emma Darwin in 1840 by George Richmond.

Additionally, there can be optional sections: an introduction that focuses on the dominant key; and a coda that may be long and elaborate, but is essentially perceived as extra material not based on the theme. This contrivance is purely for fun and not essential for your following the piece nor, hopefully, your enjoyment of it. That said, and this being the *Introduction*, I ought to get on and establish the dominant key to this piece, the role of music and musicality in Darwin's life and works.

While Charles was instrumental in changing the course of human knowledge, Emma was undoubtedly his accompanist. They were married on 29 January 1839 at St. Peter's Church, Maer in Staffordshire. Shortly afterwards, Emma's first personal piano had been gifted to her from her father, likely as a house-warming present for the new marital home; she wrote about it to her sister Elizabeth Wedgwood in February 1839, "then we went slopping through the melted snow to Broadwood's, where we tried the pianoforte and it sounded beautiful as far as we could judge [...] I like it particularly in every way, and never heard a [pianoforte] I admired more. We hope to have it home to-day" [1]. The choice was a good one. Broadwoods were a highly reputable make: in 1840, Queen Victoria married Prince Albert and they also bought a Broadwood for Buckingham Palace [3]. Emma's Broadwood cost £101 10s [2] and was one of three possible models: a Boudoir (trichord) from 1832, 7' 6" long and 3' 10" wide, it spanned 6 octaves from F to F; the similar Semi also from 1832, a little smaller at 6' 11" by 3' 9"; or the larger Bichorda produced in 1836, 7' 8" by 4' 1" with a range of 6½ octaves from C to G [3].

This piano and a later Broadwood were to feature almost daily in the following 43 years of their marriage, in London and at Downe. The pianos would provide not only a much appreciated focus for recreation in the evenings and an instrument to be taught to the children, but they were also an experimental apparatus for Darwin's work, and undoubtedly a strong stimulus for his ideas on the evolution of musicality and sexual selection.

Exposition: House Music

The newlyweds' London home was established over Christmas 1838 at 12, Upper Gower Street, tucked between Regent's Park and Russell Square, described by Francis Darwin as, "a small common-place London house, with a drawing-room in the front, and a small room behind, in which they lived for the sake of quietness. In later years my father used to laugh over the surpassing ugliness of the furniture, carpets &c., of the Gower Street house. The only redeeming feature was a better garden than most London houses have, a strip as wide as the house, and thirty yards long. Even this small space of dingy grass made their London house more tolerable to its two country-bred inhabitants" [4]. With similar sentiment, Francis' sister Henrietta gives a little more detail on the the Broadwood piano, plus a tantalising hint of future instruments, "I remember it well in its handsome mahogany case; it kept its beauty of tone longer than any later piano. For the sake of quiet they lived, grand piano and all, in the smallish back room looking on the garden, which smoky though it was, was a great boon to their country souls" [5].

Emma's piano accompanied them to Down House when they moved there in 1842. Down House is set in a 16-acre estate on the edge of the village of Downe, being what you might call a country villa. It was built in the eighteenth century and extended in the nineteenth: a three-storey bay on the garden front in 1843, a new drawing room in 1858, a verandah in 1872 and a new study at the northern end of the house in 1876. However, in honour of Darwin's great achievements accomplished therein, the old study has since been reinstated and reconstructed.

The drawing room renovation coincided with purchase of a new Broadwood. On 22nd

February 1858 Emma recorded in her diary, "[pianoforte] chosen", and five days later Darwin announces, "A grand new Pianoforte has just arrived". It was a 'Patent Repetition Bichorda Grand' Broadwood (6' 8" by 4' 4", 6¾ octaves from C to A), dating from 1854. This Wornum & Sons piece cost the Darwins £75 5s. If purchasing a new piano just prior to publication of *The Origin...* [9] seems coincidental, then even more so is the purchase of an Érard, Franz Liszt's (1811–1886) favoured make [5] and tour sponsor, in May 1870 for £141, similarly preempting *The Descent of Man* [6]. This Érard was captured in an inventory of 1882 [7], but how it was put to use remains unknown.

The pianos supposedly remained at Down House until Emma's death in 1896 when the family cleared all belongings and sold the property. Between 1907 and 1922, the house was used as a girl's boarding school, but was bought by surgeon Sir George Buckston Browne (1850-1945) in 1927 and shortly after handed over to the British Association for the Advancement of Science (BAAS) to be preserved as a Darwin museum, and this opened in June, 1929. Meanwhile, the newest Broadwood had been sold, and so the new museum had to trace it and buy it back for £20 from the Church of Humanity, formally the Positivist Society [8], who had it in their church at 19, Chapel St, London WC1. It has remained at Downe ever since, passed through the hands of The Royal College of Surgeons and the Natural History Museum and now conserved by English Heritage, meaning that the piano now displayed in the Down House drawing-room is senior to the 150-year old *On The Origin of Species* [9], by 5 years.

Emma's Broadwood in the drawing room of Down House, as it can still be seen, and is sometimes played, today.

David Stradling [10] had an opportunity to experience Emma's Broadwood close at hand,

> On a family holiday we stayed in a friends house [...] my daughter was 13 maybe 14 [...] I wanted to go as a biologist to visit Down House [...] it was very quiet and low-key [...] We sort of split up a bit looking around, and I found myself standing looking at the study [...] my daughter and wife and son were apparently in the drawing room and the caretaker [...] came up to me and said 'your daughter says you would like to play the piano?' [...] what had apparently happened is that my daughter, who is a very keen string player, had spotted a music stand in the far corner of the living room and she was craning her head to try and see what the music was on the stand and in so doing upset some sort of barrier [...] and it set off an alarm. So the caretaker came [...] and said ' Don't worry my dear, people do it ten times a day', [...] he said while it's switched off, go in there and have a look [...] and on her way back she came via the piano [...] and while she was stood there the caretaker said 'Would you like to play it?'. And she said 'Oh no, I can't play the piano, but my Dad would!' [...] I was so glad to see on the stand was a book of Schubert, which is rather a favorite of mine. So I [...] wound up playing for over half an hour. And the caretaker said 'Oh, thank you so much, it never gets played except once a year when my niece comes to stay' [...] I must admit that while I was playing there was all sort of things going through my mind, for instance, what if he was sat over there? [...] Then my wife came and said that they'd like to go!

The alarmed ropes must be a recent introduction as, amazingly, there were no such barriers, but even more liberties extended to a sixth-form school party in 1964 [11],

> The caretaker at the house was very helpful, and allowed us virtually the freedom of the house. Several of the pupils, for example, were allowed to play Emma Darwin's piano, as she used to do when Charles Darwin lay on the couch next to her. I for one was struck by the wonderful atmosphere this piano-playing created. The piano and the couch were situated in the drawing-room, and all the furniture there, as in all the other rooms we visited, was that used by Darwin.

Development: I Tunes

Leonard Darwin gives us a glimpse into those restful evenings in Down House, and a privileged insight into his father's humble personality [12],

> I wish I could paint in words a picture of my father lying quietly on the sofa in the drawing-room, whilst my mother was playing, and playing beautifully, some slow movement of Beethoven. Little was said, but I am sure that the music was not without effect on my father's mind. And if what he had thus gained had gone out of his mind when he was writing his autobiography, the explanation is to be found in the modesty of his nature, which led him to concentrate his attention on possible defects in his own character and to ignore probable merits.

This modesty bound Darwin to admit in his autobiography that he had always been very unmusical; The running joke was that he had such a bad ear that evidently he needed nudging to stand when the National Anthem was playing, simply because he failed to recognise it, rather than any nonconformist anti-royalism. During his early days in Edinburgh, he recalled [2],

> I also got into a musical set [...] From associating with these men, and hearing them play, I acquired a strong taste for music, and used very often to time my walks so as to hear on week days the anthem in King's College Chapel. This gave me intense pleasure, so that my backbone would sometimes shiver [...] Nevertheless I am so utterly destitute of an

> ear, that I cannot perceive a discord, or keep time and hum a tune correctly; and it is a mystery how I could possibly have derived pleasure from music [...] My musical friends soon perceived my state, and sometimes amused themselves by making me pass an examination, which consisted in ascertaining how many tunes I could recognise, when they were played rather more quickly or slowly than usual. 'God save the King,' when thus played, was a sore puzzle.

And this could be corroborated by one of the instigators of these pranks, "What gave him the greatest delight was some grand symphony or overture, of Mozart's or Beethoven's, with their full harmonies. For simple melody he cared little, and indeed he was so deficient in the power of distinguishing tune or remembering it, that he cd. not recognize *God save the King*, or any of the most popular airs of the day when played to him on a flute" [13].

Many years later, one visitor to the house discovered that, with the passing of time and in a Lamarkian sense, Darwin's enthusiasm for music may have dwindled from those earlier days [14],

> After the family's late dinner was over it was Mr. Darwin's habit to take a little rest, and then to join the family in the drawing room. Mrs. Darwin and several of the children were very fond of music, and the grand piano was in very frequent use. I have often watched Mr. Darwin sitting by, with an inquiring expression on his face, as a sonata of Mozart or Beethoven, or a nocturne of Chopin's was being played.
>
> I remember well his saying, "When I was young and after we had been for some time married, I took great pleasure in hearing music, and I was a good listener. But by degrees, and as I became more and more absorbed in scientific observation, the taste for music lessened and fell off. I really believe the musical department of my brain became atrophied from want of use. I have often lamented it since, but it has gone too far to recall now. Depend upon it, it's a great mistake to lose interest in any art or hobby by disuse.

"After dinner ... he would often lie on the sofa and listen to my mother play the piano" [2].

However, perhaps because of Darwin's self-deprecation, Francis felt it necessary to defend his father's musical tastes [15],

> He used to lament that his love of music had become dimmed with age — yet in my recollection, his love of a good tune was strong. I never heard him hum more than one tune — *Ar hyd a nos* [All Through The Night] — which he went through correctly — he used too, I believe, to hum a little Otaheiti song. From his want of ear he was unable to recognize a tune when he heard it again; but he remained constant to the tunes he liked and would say, "That's a fine thing (or a good tune?)." "What is it?" He liked especially bits of Beethoven's symphonies and bits of Handel. He enjoyed Effie's [Katherine Euphemia Wedgwood was Fanny and Hensleigh Wedgwood's daughter] singing extremely and was much touched by such songs as Sullivan's "Will he come." He was pathetically humble about his own taste and seemed pleased when he found others agreed with him.

Nonetheless, in his own writing, Darwin was even more likely to shy behind the modesty that Leonard touched on. Writing to Emma from "The Mount", the family home in Shrewsbury, while visiting his 82-year-old father of failing health for the last time in May 1848, Darwin clearly considers musicality an inherited trait, applauding his wife's musicianship, and slating his own lack, "My father kept pretty well all yesterday, [...] Thanks for your very nice letter received this morning, with all the news about the dear children: I suppose now and be-hanged to you, you will allow Annie is "something." I believe as Sir J. L. said of his friend, that she is a second Mozart; anyhow she is more than a Mozart considering her Darwin blood" [16]. Here he was alluding to that most celebrated of wunderkinder, who perhaps due to continental foppery, seems to be from a bygone age of powdered wigs, breeches and brightly-coloured silk jackets, unlike the dowdy Victorian era. Actually, he would have been only 53 years old in Darwin's birth year if he had not conformed to life expectancy in central Europe for 1791 [17] and died aged 35. Perhaps why he felt Mozart was a topical reference to be employed elsewhere, in *The Origin...*, by which to clarify which elements of musicality are inherited [8],

> If we suppose any habitual action to become inherited-and I think it can be shown that this does sometimes happen-then the resemblance between what originally was a habit and an instinct becomes so close as not to be distinguished. If Mozart, instead of playing the pianoforte at three years old with wonderfully little practice, had played a tune with no practice at all, he might truly be said to have done so instinctively. But it would be the most serious error to suppose that the greater number of instincts have been acquired by habit in one generation, and then transmitted by inheritance to succeeding generations [18].

Recapitulation: Experimental Music

A self-confessed lack of musicality did not stop Darwin investigating the adaptive benefits in those better equipped than himself. A famous example [19] suggests one possible use of the cheaper Érard piano, if Emma's Broadwood was considered too precious,

> Worms do not possess any sense of hearing. They took not the least notice of the shrill notes from a metal whistle, which was repeatedly sounded near them; nor did they of the

> deepest and loudest tones of a bassoon. They were indifferent to shouts, if care was taken that the breath did not strike them. When placed on a table close to the keys of a piano, which was played as loudly as possible, they remained perfectly quiet.

Darwin was clearly interested in the earthworms' capacity to sense sound, but that wasn't all that he tried. He also tested their response to smoke, light, heat, vibration, touch, smell, and even different tastes. From the apparent arbitrariness of his experiments, Darwin was able to prove that [19],

> Worms are poorly provided with sense-organs, for they cannot be said to see, although they can just distinguish between light and darkness; they are completely deaf, and have only a feeble power of smell; the sense of touch alone is well developed. They can therefore learn little about the outside world, and it is surprising that they should exhibit some skill in lining their burrows with their castings and with leaves, and in the case of some species in piling up their castings into tower-like constructions. But it is far more surprising that they should apparently exhibit some degree of intelligence instead of a mere blind instinctive impulse, in their manner of plugging up the mouths of their burrows.

Yet these conclusions could not be drawn from any random set of observations, so what informed Darwin in his experimental design? After all that has been said on Darwin, we are still left ignorant of his working mind. Perhaps the closest we will ever get to an explanation are within a few notes sketched by Francis [15], writing of his recently deceased father's use of instinct in his work and about some unconventional musical requests,

> He often said that no one could be a good observer unless they were an active theoriser. This brings one back to what I said about his instinct for exceptions — it was as tho' he were charged with theorising power ready to flow into any channel on the slightest disturbance — so that no fact however small could avoid releasing some of this theory — & thus becoming magnified into strong perceptibility. In this way it naturally happened that many untenable theories occurred to him — but fortunately his richness of imagination was equalled by his power of judging & condemning the thoughts that occurred to him. But he was just to his mind theories, and did not condemn them unheard; and so it happened that he was willing to test what would seem to most people

> not at all worth testing. These rather wild trials he called "fool's experiments" & enjoyed extremely. As an example of a fool's experiment he found the cotyledons of *Biophytum* so sensitive to vibration of the table &c. that it occurred to him that they might perceive the vibrations of sound, & therefore he got me to play my bassoon close to a plant [...] his wish to test the most improbable ideas. This wish was very strong in him and I can remember the way he said "I shan't be easy till I've tried it" — as if an outside force were compelling him.

"He was willing to test what would seem to most people not at all worth testing". Therein lies the crux. Willingness. Willingness because he wasn't shackled by prejudice and procedure, but was free to improvise.

Other musical notes played throughout Darwin's work. Ultimately, he saw functional structures for sound production as an adaptive, and thus inheritable, trait while the appreciation of music acts as a medium for sexual selection, "unless the females were able to appreciate such sounds and were excited or charmed by them, the persevering efforts of the males, and the complex structures often possessed by them alone, would be useless; and this it is impossible to believe" [Error: Reference source not found].

Darwin saw a natural progression from sound production through to language development, such that expressive sounds would have arisen from an ancient vocabulary [Error: Reference source not found],

> Even monkeys express strong feelings in different tones--anger and impatience by low,-- fear and pain by high notes [...] The sensations and ideas thus excited in us by music, or expressed by the cadences of oratory, appear from their vagueness, yet depth, like mental reversions to the emotions and thoughts of a long-past age [...] if we were to admit that man's musical capacity has been developed from the tones used in impassioned speech. We must suppose that the rhythms and cadences of oratory are derived from previously developed musical powers [...] I conclude that musical notes and rhythm were first acquired by the male or female progenitors of mankind for the sake of charming the opposite sex [...] it appears probable that the progenitors of man, either the males or females or both sexes, before acquiring the power of expressing their mutual love in articulate language, endeavoured to charm each other with musical notes and rhythm.

A preference-for-music motivated sexual selection can therefore account for the beginnings of language and song, their culmination in the likes of Shakespeare and Mozart, and everything that has transpired since, from Albinoni to ZZ Top, and beyond.

Coda

Emma's partnership and musicality played a perfect accompaniment to Darwin's day. The fates of her pianos provide a compelling conundrum.

Of the Broadwoods, there is an additional twist. Henrietta also tells us that in 1873, "Elizabeth Wedgwood's sight had been failing more and more for some time, a privation she bore with the utmost patience. But my mother used often to say how sad she felt it to come in and find her doing nothing, when her life had been one of continual activity. My mother gave her her old Broadwood grand-piano, and to fill up some of her weary useless time helped her to learn by heart simple airs to play to herself" [5]. If this was the piano from Upper Gower St., then could there have been three pianos about Down House for a few years until the old Broadwood was given to Elizabeth? Assuming that the old Broadwood was never returned following Elizabeth's death on 7th November 1880 [20], the new Broadwood and the Érard were likely sold on in 1896, perhaps with the house. By 1929 the new Broadwood had returned, but what had become of the Érard is a mystery which has not been pieced together.

Sadly, the Downe collection also has no surviving sheet music nor instruments [21], such as Francis' bassoon. The only other piece of relevant music furniture in Down House today, included in the 1882 inventory, is a Victorian Walnut Music Canterbury (mainly used for storing sheet music), leaving Emma's Broadwood to take centre stage.

Acknowledgements

Fortississimo thanks to: Randal Keynes for all his excellent work at Downe, and for helping unravel Darwin's personality and family life; Cathy Power and Annie Kemkaran-Smith for looking into the Down House archives; Finchcocks Musical Museum for advice on piano makes; Henry Nicholls and Peter Arnott for their research of the "Mendel." story; Fiona Alexander, violinist with the Scottish Chamber Orchestra, for proof reading with a musician's eye; and David Stradling for his Down House anecdote.

Notes and Bibliography

1. Henrietta Litchfield, ed. 1915. *Emma Darwin, A century of family letters, 1792-1896*. London: John Murray. Volume 2.

2. An indication of the pianoforte's increasing popularity, and Broadwood's reputation for quality instruments, by 1848 a Broadwood grand in rosewood cost 155 guineas. David Wainright (1982) *Broadwood by Appointment: A History*. Quiller Press.

3. John Broadwood & Sons. Available http://www.uk-piano.org/broadwood/.

4. Francis Darwin, ed. 1887. *The life and letters of Charles Darwin, including an autobiographical chapter.* London: John Murray.

5. Program notes: Paris in the 19th Century, Sunday, October 3 2004, 3 pm, Chan Centre, UBC.

6. Charles Darwin 1871. *The Descent of Man, and Selection in Relation to Sex*. John Murray.

7. Annie Kemkaran-Smith, English Heritage, *pers. comm.*

8. Vassos Argyrou 2005. *The Logic of Environmentalism: Anthropology, Ecology and Postcoloniality*. Berghahn Books.

9. Charles Darwin 1859. *On the Origin of Species by Means of Natural Selection, or the Preservation of Favoured Races in the Struggle for Life*. John Murray.

10. David Stradling is Chairman of Whitley Wildlife Conservation Trust at Paignton Zoo, following a long career in ecological entomology, in particular neotropical leaf-cutting ants, as well as tropical tarantula spiders. He was elected a Fellow of the Royal Entomological Society in 1970 and has worked for the Nature Conservancy in Wareham; The Commonwealth Institute of Biological Control in Europe and South America; the University of Bristol; Long Ashton Horticultural Research Station, the University of West Indies and Exeter University, as well as advising on several wildlife films.

11. Julie Earl 1964. A Visit To Down House. *Phoenix* (Drayton Manor Grammar School, December, 1964), 10. Available http://www.dmgs67.co.uk/phoenix.htm.

12. Leonard Darwin 1929. Memories of Down House. *The Nineteenth Century* 106:118-123.

13. John Maurice Herbert, 2 June 1882. [Recollections of Darwin at Cambridge]. CUL-DAR112.B57-B76.

14. Wallis Nash 1919. *A lawyer's life on two continents*. Boston: Richard G. Badger, the Gorham Press.

15. Francis Darwin c.1884. [Preliminary draft of] *Reminiscences of My Father's Everyday Life*. CUL-DAR140.3.1 —159.

16. Letter 1176 — Darwin, C. R. to Darwin, Emma, [20–1 May 1848].

17. Mikuláš Teich & Roy Porter 1996. *The Industrial Revolution in National Context: Europe and the USA*. Cambridge University Press.

18. On the subject of inheritance, there is a wonderfully tantalising entry in Emma Darwin's diary for Monday, 26th November, 1866 (Available http://darwin-online.org.uk/EmmaDiaries.html) which reads "Mendel. ottett !! Wilhelmj Pop. concert". It has been suggested that Gregor Mendel could have sent copies of his work to Darwin because he greatly admired him, but none, nor record of any, has ever been found amongst Darwin's library (Nino Strachey, former Curator, Down House, cited in 'Sclater 2006. The extent of Charles Darwin's knowledge of Mendel. *Journal of Bioscience* 31: 191-3'). But, does Emma's diary full of shorthand and codes suggest that they actually met? Alas, no. That evening they were to simply enjoy a concert of Mendelssohn's *Octet in E-flat major, Op. 20*, about which Emma was clearly excited using double exclamation marks. This series of "Monday Popular Concerts" was particularly auspicious. Advertised in *The Times* and attended by Prince Leopold, it to be held at London's principal concert venue, St. James' Hall. It was also the first English tour for the 21-year old violin soloist, a prodigy hailed as the "German Paganini", August Daniel Ferdinand Victor Wilhelmj (1845-1908). The concert was also to include Beethoven's *Romance in F*, and Mendelssohn's *C Minor Trio* performed with Charles Hallé (pianoforte) and Piatti (violoncello) (*The Musical Times and Singing Class Circular*, Vol. 42, No. 700, Jun. 1, 1901, pp. 369-374. Musical Times Publications Ltd.). The tour was a sensation. The only criticism from the press, levied by the *The Morning Post*, suggested an alternative spelling for the violinist, "Wilhelmi would be more simple and intelligible to English eyes".

19. Charles Darwin 1881. The formation of vegetable mould, through the action of worms, with observations on their habits. London: John Murray.

20. Edna Healey 2001. *Emma Darwin: The Inspirational Wife of a Genius*. Headline Book Publishing.

21. Cathy Power, English Heritage, *pers. comm.*

Rich Pickings

Alas, the mudslinging continues as once more Richard Dawkins attracts vitriol and his defenders retaliate, and yet the proverbial barn side remains intact and mud free. The great shame is that when engaging in rational debate, it is no place for hotheaded name calling; arguments have to be securely anchored in truth, there has to be some substance to those mud pies. The evidential documentation is out there, but people seem too ready to write before they read. Darwin would never have condoned such sloppiness.

In his *Guardian* article *Natural selection: give me Darwin over Dawkins any day*, Jonathan Jones claims that, "[Charles Darwin] was not self-consciously clever: he never talked down to his readers. His masterpiece, *On the Origin of Species*, is a modest book" and uses the line to attack Richard Dawkins, whom he slights as just wanting, "to be the cleverest kid in the class"

In so doing, Jones uses Darwin's apparent legendary shyness and self-doubt to offset Richard Dawkins' palpable superciliousness and self-importance.

In response, Alan Henness defends Dawkins against Jones' accusations of arrogance and lack of empirical evidence in his skeptic-based *Twenty-First Floor* article *Being A Dick To Dick?*, by drawing comparisons between Victorian sensibility and modern day boldness, thereby casting Dawkins as a latter day champion: "How do you compare Darwin with Dawkins? You don't. They are different people in different times. The Victorians needed Darwin; we need Dawkins".

Of course, both Jones and Henness are impaired by their lack of authority: Jones erects a towering straw figure of tolerance and understanding, and Henness fails to knock it down. With the incredible resources that we now have readily available via *Darwin Online* and *The Darwin Correspondence Project,* they really have no excuse in not being better informed. Consulting those invaluable resources it is quickly evident that if ever Darwin was a patient man, he clearly becomes less so with age, displaying much acrimony later in life towards having to deal with ignorant corespondents.

Here follows a small selection of examples that illustrate this very point, from the publication of the *On the Origin of Species* onwards, the modest masterpiece alluded to by Jones. But, please don't be satisfied with only these, take yourself off to the archives and locate many more.

Meanwhile, picking up the story soon after the first edition, the publisher John Murray was repeatedly asking a nonplussed Darwin for a new edition of *The Origin*. This was a welcome opportunity for Darwin to make a number of corrections and additions,

> *in the hopes of making my many rather stupid reviewers at least understand what is meant. I hope & think I shall improve the Book considerably (CD letter to John Murray, 5 December [1860]).*

A year later, Darwin writes to Charles Lyell recounting notification via Asa Gray that Frances Bowen (Professor of Philosophy at Harvard University) and Louis Agassiz (Professor of Natural History at Harvard) were opposing Darwin's ideas on transmutation. Darwin is incredulous,

> *I cannot understand what Agassiz is driving at. You once spoke, I think, of Prof. Bowen, as a very clever man. I shd have thought him a singularly unobservant & weak man from his writings. If ever he agrees with me on any one point, I shall conclude that I must be in error on that. He never can have seen much of animals or he would seen the difference of old & wise dogs & young ones. His paper about hereditariness beats everything. Tell a breeder that he might pick out his worst individual animals & breed from them & hope to win a prize; & he would think you not a fool, but insane. I believe Bowen is a metaphysician & that I presume accounts for an entire want of common sense (CD letter to Asa Gray, 11 Apr [1861]).*

Bowen published two reviews of The Origin in 1860, the second Darwin dismissed as being, "absurd", "monstrous" and "rubbish" (CD letter to Asa Gray, 26 Nov [1860]).

When Thomas Henry Huxley delivered two lectures at the Philosophical Institution of Edinburgh, on man's relation to the lower animals, he received fierce criticism in an article that appeared in the Presbyterian newspaper, the Witness, on 14 January 1862. Huxley sent Darwin a copy of the article. He responded,

> *I have been much amused at the Witness. Such abuse is as good as praise. What fools these Bigots are (CD letter to T.H. Huxley, 22 Jan [1862]).*

Darwin even went on to show that as long as the source of information is trustworthy, a fool is a fool regardless of remoteness; he is openly disparaging of a third party involved in a quarrel that he was not directly involved in,

> *You have given excellent counsel to Bates & I hope he will follow it; what an old malignant fool Dr Grey is; but I never care an atom for his malignacy; it never makes me angry, & I believe your explanation is right; one gets used to it (CD letter to J.D. Hooker, 15 & 22 May [1863]).*

Clearly the form that idiocy and ignorance presented itself was unimportant to Darwin. Being house-bound for long periods, his correspondence was a life-line to the outside world and he expertly choreographed a staggering network of associates. So, when time-wasters presented themselves, he was understandably annoyed by those, "bothering correspondents" who "seem to increase in number and in folly". However, ever the gentleman, he apparently honored their invitation for social contract, albeit belligerently, moaning afterwards of having "just answered two precious fools" (CD letter to Leonard Darwin, 25 Nov [1874]), and complaining that, "half the fools throughout Europe write to ask me the stupidest questions" (CD letter to Reginald Darwin, 8 Apr [1879]).

However, there were particular individuals who came to test his patience to the limit. In fact, the very example that sprung to mind on reading Henness' response to Jones was Darwin's explosive exasperation with Frederic William Farrar, originally a schoolteacher and clergyman, who was promoted to public school master and broad churchman, before becoming Canon of Westminster from 1875. Farrar interest in linguistics led him to write a book Chapters on Language that based language development on imitation, a copy of which he sent to Darwin. Farrar's book also suggested that Darwin's ideas had not conclusively linked man to animals. Darwin was unimpressed with Farrar's evident lack of knowledge on the subject, and perhaps more directly than with any other individual, he let his feeling be known. Thus, given their relationship, and Darwin's thoughts on the man, it is ironic that Farrar was in a position to preach Darwin's funeral sermon.

Darwin initially acknowledged receipt of Farrar's book,

Down. | Bromley. | Kent. S.E.

Wed. Oct 11

Dear Sir

I am very much obliged to you for your kind present of your work on Language.

No subject in my opinion can be more interesting, & I hope soon to read your book, but I am not sure that I shall be able as my health is at present weak.

From some of your papers which I have read I feel certain that your opinions on all subjects would be most liberal & fair.

With my best thanks | I remain yours faithfully | Ch. Darwin

(CD letter to F.W. Farrar, 11 Oct [1865])

Later he managed to read it and form a response,

Dear Sir

As I have never studied the science of language it may perhaps be presumptuous, but I cannot resist the pleasure of telling you what interest & pleasure I have derived from hearing read aloud your volume.

I formerly read Max Müller & thought his theory (if it deserves to be called so) both obscure & weak; & now after hearing what you say, I feel sure that this is the case & that your cause will ultimately triumph.

My indirect interest in your book has been increased from Mr Hensleigh Wedgwood, whom you often quote, being my brother in law.

No one could dissent from my views on the modification of species with more courtesy than you do. But from the tenor of your mind I feel an entire & comfortable conviction (& which cannot possibly be disturbed) that if your studies led you to attend much to general questions in Natural History, you wd come to the same conclusions that I have done.

Have you ever read Huxley's little book of Six Lectures   I wd gladly send you a copy if you think you would read it.

Considering what Geology teaches us, the argument for the supposed [my emphasis] immutability of specific Types seems to me much the same as if, in a nation which had no old writings, some wise old savage was to say that his language had never changed; but my metaphor is too long to fill up.

Pray believe me dear Sir yours very sincerely obliged | Ch. Darwin

(CD letter to F.W. Farrar, 2 Nov [1865])

The "supposed" here, and Darwin's succinct repudiation is scalding of Farrar's supposition in the absence of evidence. Here Darwin is clearly calling for informed debate by not 'suffering fools gladly'. Despite this, the most beautiful and telling set of correspondence regarding Darwin's belief in informed and independent thought, fueled through self-education and reading, comes from a short series of letters as a result of an invitation received from the Archbishop of Canterbury's office.

38, Belgrave Road. S. W.

Nov. 2

Dear Sir

On Jan 7th last an informal conference was held at Lambeth Palace of Scientific men; the Archibishop of Canterbury in the Chair, when it was resolved unanimously, "That it is desirable that those Scientific men who believe in the truths of Religion should take every opportunity of stating that belief; and that the following be appointed a Committee with power to add to their number, for the purpose of maintaining communication among those desirous of promoting this object: His Grace the Archibishop of Canterbury, Prof[ess]or Stokes Prof[ess]or Balfour Stewart, Dr Sorby, Dr Gladstone, Prof[ess]or Rollestone and Prof[ess]or McKendrick, Hon. Sec Mr Walter Browne". The Committee have lately been able to take advantage of an offer, kindly made to them by the proprietors of the "Contemporary Review', who propose to publish a series of articles dealing with the present state of the various Sciences. It is proposed to consider how far the theories in each science, without any reference to Christianity, rest on fully proved & verified laws, & how far on hypotheses, conjectures more or less probable; in other words how far each science has advanced [...] I am requested by the Committee to inquire whether you would kindly undertake to prepare an article of this character on the Science of Comparative Anatomy. The points which it is suggested, should be specially dealt with, are the evidence on the Theory of Evolution; but of course, much would be left to your own discretion in this matter. It is proposed that the series should commence as early next year as possible and I may add that Dr Gladstone, Dr Sorby & Prof[ess]or Balfour Stewart have already consented to take part in this. The honorarium for each article will be £1 per page & the article should run to about twenty pages. It is hope that the series may be subsequently published in a volume.

Hoping for an early & favourable reply | I remain | Yours truly | Walter R Browne | Hon. Sec

(W.R. Browne letter to CD, Nov [1880]).

Darwin tactfully responded,

Sir

The state of my health will not allow me to attend the meeting at Lambeth Palace, though I should feel it an honour to meet there so many distinguished men. It would, however, not be sincere on my part to assign want of strength as the sole reason for not attending, in as much as I can see no prospect of any benefit arising from the proposed conference.

I beg leave to remain, Sir, | Your obedient servant | Charles Darwin

(CD letter to W.R. Browne, 18 Dec [1880]).

But Browne pursued the now-famous naturalist for any input possible,

Dear Sir,

I regret to learn that your health will in any case prevent your attending the proposed Conference at Lambeth: but to prevent your having any misconception as to its objects, I venture to enclose a proof of the Proceeding as at present sketched out. I know the [Archbishop] is anxious to obtain expressions of opinion, even from those who do not see the desirability of holding such a meeting; & that he wd particularly value such an expression from you. I am sure therefore he wd be obliged if, in returning the paper, you could kindly make any remarks either on particular points, or on the subject in general.

Yours truly | Walter R Browne

(W.R. Browne letter to CD, 21 Dec [1880]).

To which Darwin quickly replies, proposing that every one has the capacity and right to study the evidence, and form their own conclusions.

Down, | Beckenham, Kent. | (Railway Station | Orpington. S.E.R.)

Dec 22 1880

Dear Sir

I am much obliged for your very courteous note. I regret that it wd be impossible for me to explain the causes of my disbelief in any good being derived from the conference, without treating the subject at inordinate length. I will only add that in my opinion, a man who wishes to form a judgment on this subject, must weigh the evidence for himself;

& he ought not to be influenced by being told that a considerable number of scientific men can reconcile the results of science with revealed or or natural religion, whilst others cannot do so.

I beg leave to remain | Dear Sir | Yours faithfully | Charles Darwin

(CD letter to W.R. Browne, 22 Dec [1880]).

It is clear from his correspondence and autobiographical pieces that Charles Darwin set his own standards for study; he sought the evidence that he required to form informed opinion. But, importantly, if others wished to engage in debate with him, then he also expected the same of them. This is the same requirement that Dawkins and other skeptics, rationalists, atheists and neo-Darwinists, should be requesting, and typically do so; in contrast, it is the notable lack of evidence why psychics are so easily debunked, the arguments of religious apologists are fatuous, and Intelligent Design advocates are finding no footing in scientific debate. Moreover, it is sad to see that even people nowadays discussing Darwin himself, his character and attitudes, and doing so in high profile public fora, are also failing to live up to Darwin's expectations.

Darwin in Disguise

The many guises of Darwinism

One does not need to step outside the scientific arena to encounter a bitter and aggressive battle over Darwinism. The argument centres on the extrapolation of Darwinian mechanisms beyond evolutionary boundaries, where there has been no shortage of new disciplines devised: Neural Darwinism, Darwinian Medicine, Evolutionary Psychology, Evolutionary Epistemology, Evolutionary Ethics, Evolutionary Computation, Evolutionary Cosmology, Memetics, Digital Darwinism, Corporate Darwinism, Universal Darwinism, Quantum Darwinism and Social Darwinism. The struggle to make natural selection fit has been going on ever since the publication of *The Origin of Species* [1] on 24th, November 1859.

The fields of study listed above have taken one, or more, aspects of Darwinian evolution and fused them with existing paradigms, in each case creating a new discipline. This has been with varying degrees of success. On the anniversaries of Darwin's birth and publication of *The Origin of Species*, we might pause to assess this practise, and our treatment of his ideas. This article collects together comments from people working in a range of disciplines in an attempt to assess the ongoing influence of Darwinism beyond the historical context of evolutionary biology. From each participant I requested an informal account of the impact of Darwin on modern thought, which could also be a personal account of the impact of Darwin on the individual's thought. Not surprisingly, many participants referred to those influences within their own fields, yet the passion with which Darwinism was defended, and personal influences were recounted, was surprising.

Richard C. Lewontin has been a long-term advocate of a multi-disciplinary approach for studying human nature, and a staunch critic of gene-oriented reductionism in the social sciences,

> The problem is that many have turned Darwin's description of the way in which organic evolution works into (1) a speculative tool for inventing a natural selective explanation for everything in the world, with no conceivable way of checking on the reality of these "Just So" stories and (2) have extended Darwin's structure which was tied to a particular natural phenomenology- the biological reproduction of offspring and the differential probability of survivorship and reproduction of those offspring in a real world of biological objects into a generalized metaphor for every kind of historical change in human culture and human history. This has led to the production of a vast literature on sociobiology, "evolutionary" psychology, "evolutionary" accounts of history, "evolution" of culture which are all intellectual games that vulgarize the Darwinian explanatory structure in the interest of producing a general theory of everything.
>
> As Norbert Weiner once said: "The price of metaphor is eternal vigilance".

We shall now take a look at seven key ideas to illustrate the broad range of ways in which Lewontin considers Darwinism to have been misappropriated. We will begin with perhaps the most extreme example, evolutionary cosmology.

Evolutionary Cosmology

Lee Smolin, a theoretical physicist at the University of Waterloo, advanced the concept of Evolutionary Cosmology [*e.g.*, 2], which proposes that an evolutionary mechanism underlies universe survival. If energy fluctuations in the parental "quantum foam", the vacuum precursor of universes, exceed a certain threshold, then an expanding, persistent bubble universe forms. Otherwise a small, temporary universe blips and dies. Bubble universes like our own that do survive, form matter and galactic structures, and can even propagate their own bubble children through the collapse of black holes. It follows that the more black holes a "fecund universe" contains, the more offspring it can spawn, and the longer it will persist. This gives rise to variation between universes, and a measure of reproductive success upon which selection can operate, and these are the prerequisites of Darwinian evolution.

Rupert Sheldrake reveres Darwin among four in a personal hall of fame, alongside Goethe, Wallace and Frederic Myers, citing *The Origin of Species* and *The Expression of the Emotions in Man and Animals* [3] as having had a huge influence on his thinking. Sheldrake came to prominence for his hypothesis on morphic resonance, which, he explains, "is really based on taking an evolutionary principle beyond the realms of biology and applying it to cosmology. Of course, in Darwin's time people thought that the cosmos didn't evolve, only life did, but since the 1960's we've had an evolutionary cosmology, so the evolutionary paradigm has now become dominant over all of science".

> My work on morphic resonance is based on the idea that the so called laws of nature themselves evolve. What we've got at the moment is an unsteady compromise between evolutionary thinking and old style physical thinking in terms of fixed laws, eternal laws of nature that have been in place since the moment of the big bang. So physicists tend to assume, and so do all other scientists, that nature is governed by fixed laws but that the actual contents of nature evolve. And I think that the laws themselves may evolve and, in fact, the word "law" is very inappropriate, I think they're more like habits. Nature is governed by evolving habits [...]
>
> I think all self-organising systems, including atoms, molecules, crystals, living things of all kinds, social groups, have organising morphic fields which have a kind of a inherent memory. Each species has a collective memory to which individuals contribute and on which they draw. The same is true of crystals and molecules: they have a kind of memory too and that the so-called "laws" which govern molecules and crystals and most self-organising phenomena of nature are not really laws at all, they're habits [...]
>
> I propose that there is a memory in all of nature and that the natural regularities depend on these habits. All sorts of evolution depend on the interplay between habits and creativity. Habits give repetition, creativity throws up new variations, and natural selection selects among new forms of plants or animals or new patterns of behaviour or whatever, and those that are repeated become increasingly habitual because of morphic resonance.

So, as you will have noticed, this article is all about scientific ideas. Some might say unorthodox scientific ideas. Their unorthodoxy comes from their acceptance, or lack thereof, within the wider scientific community; as one knows, in science, ideas must

remain hypothetical until proof can be found. Exactly how much proof is required is a different matter. Nonetheless, ideas prove highly popular before fully-well founded. Not least, ideas that touch on the core of our humanity and individuality, our brain function and its products. Given our inalienable pigeon-holing, it is perhaps less than surprising that we must even categorise ideas about having ideas.

Memetics

In the eleventh chapter of *The Selfish Gene* [4], Richard Dawkins famously introduced the concept of the meme. There he skilfully crafted a concept that took neo-Darwinian ideas and applied them to culture: "Cultural transmission is analogous to genetic transmission in that, although basically conservative, it can give rise to a form of evolution [...] We need a name for the new replicator, a noun that conveys the idea of a unit of cultural transmission [...] *meme* [...] pronounced to rhyme with cream" [4]. This was surprisingly acceptable to the popular science-reading public. Even more surprising because Darwinian concepts are not widely accepted amongst the general public, whereas memetics is much more so, even though memetics lacks the overwhelming evidence that exists for evolution.

Like all good metaphoric memes, memetics has enjoyed a meteoric uptake, in society if not amongst academics. Searching Google Scholar <http://scholar.google.com/> on 1st August 2008, the term "gene*" returned 2,340,000 pages, whereas "meme*" only returned 410,000. But, search the internet as a whole <http://www.google.com/> and "meme*" beat "gene*" by 45,500,000 to 28,100,000 pages even after controlling for coincidences of the same word stem, foreign languages and peoples' names.

Why is memetics so popular outside of academia? Perhaps it is because imagination-fuelled culture tends towards the larger-than-life; memes can be anything from paperclips to proverbs, marketing ads to Mozart arias. As Dawkins wrote, "Darwinism is too big a theory to be confined to the narrow context of the gene". Whatever the answer, the meme of Memetics is undeniably a most successful idea, arguably popularized most successfully by Susan Blackmore,

> I never would have written *The Meme Machine* [5] if it weren't for the fact that Richard Dawkins took Darwin's central insight about the design process, called it 'Universal Darwinism', and showed me how broadly it applies. So, I would now stick my neck out [...], even further than Richard Dawkins does, and say that this is the only design process in the universe, *i.e.*, the process of natural selection or [...] copying with variation and selection [...] Other things like self-organization do not create design for function in the same sense. The great implication of this is not just that all of biology is designed by that process, but, whenever I do something that feels like it is of my own free will, or comes from something I call "the self", or when people do creative things, like writing books, painting pictures, having new, scientific ideas, all of those, in my opinion, are examples of design by natural selection. We are not yet close to understanding how that is, but I think we will.

Irrespective of its popularity, memetics receives a lot of criticism, mainly in the form of claims that it fails as a science. For example, Luis Benítez-Bribiesca, who is Editor in Chief of the Archives of Medical Research, has asserted that, "there has been as yet no scientific demonstration of such an immaterial replicator" and so "there is no clear-cut definition of a meme" [6]. This is argued not least because memes lack a replicating code-script (*sensu* Schrödinger [7]) that can be acted on in the same way by our minds, as natural selection acts on DNA.

Sexology
Sexology is defined as the scientific study of human sex and sexuality. Alfred Kinsey was a professor of entomology and zoology specialising in gallwasps. When Kinsey became interested in different forms of human sexual practices, he transferred his

appreciation of variation in individual gallwasps to human sexual behaviour. His recognition of variation is resonant of Darwin's own considerations of species, that species are a convenient construct, but clear boundaries do not exist in nature (Darwin doesn't actually define the term "species" at any point in *The Origin of Species*). Several parallels can therefore be made between the ideas of Darwin and Alfred Kinsey, and how their ideas were received. John Bancroft, Director Emeritus of The Kinsey Institute for Research in Sex, Gender and Reproduction at Indiana University, expounds further,

> Alfred Kinsey was preoccupied with the individual variability in nature which to him seemed to make a mockery of the conventional categorizations of species, *etc*. This theme carried over into his sex research where to him the prevailing picture was of no one individual being like any other, defying meaningful categorization [...] As a sex researcher, the most important influence on my thinking has been Alfred Kinsey. Kinsey revered Charles Darwin, was clearly much influenced by him, and shared with him the experience of causing moral outrage. With his early work as a biologist, Kinsey demonstrated the extraordinary variability of one 'so called' species of insect, the gall wasp. This was the basis of his dismissal of the conventional approach to speciation. Such an approach was also very evident in his research on human sexuality, which led him to conclude that, in terms of their sexuality, no two human beings were the same. Somewhat late in my career, Darwin, via Kinsey, has had a major impact on my thinking. In my time at the Kinsey Institute we developed a systematic approach to measuring individual variability in sexual responsiveness, which has opened up a whole new research agenda. This has confronted me with the extraordinary lack of attention to individual differences in sex research that has prevailed in spite of Kinsey, and Darwin before him, and which, I believe, is evident in many other research fields. It remains to be seen how much of this variability can be attributed to genetic factors.

The similarities do not end there. Kinsey's high school classmates called him 'the next Darwin' because of his biological knowledge (he knew more than his biology teacher). On publishing his seminal works, collectively the Kinsey Reports, he was exposed to prejudices similar to those that followed publication of the *The Origin of Species*. By reporting, "shockingly high rates of non-normative sexual behavior" [8],

Kinsey's work was seen to present, "a challenge to the church" [9] and was criticized for, "no recognition of higher spiritual goals" [10]. Following publication of *Sexual Behaviour in the Human Female* [11], Kinsey answered his moralizing critics with, "The scientist who observes and describes reality is attacked as an enemy of faith. We should be able to discover more intelligent ways of protecting social interests without doing such irreparable damage to so many individuals".

Darwinian Medicine
Williams and Nesse [12] famously introduced "Darwinian Medicine" to classify an emerging idea, that humans are products of natural selection and thus, during development, our bodies will have been susceptible to compromises and limitations intrinsic to evolution. They proposed an explanation of sickness and disease in terms of the costs versus benefits to physiological function. For example, hemochromatosis might have protected against the bubonic plague, morning sickness could protect foetuses from dietary toxins, and diabetes might have helped peoples living in the northern hemisphere to survive ice ages. This new medical paradigm also diagnoses some conditions in a more positive light; pain, nausea, vomiting, diarrhoea, fatigue and anxiety are now recognised as bodily defences, suggesting that their treatment should be less interventional. Lewis Wolpert thinks Darwinian evolution is the framework for both practise and theory in medicine,

> Darwin has played an important role in my scientific life. Most significant was my recognition of Darwinian Medicine which I taught to small groups of first year medical students and was greatly influenced by *Why We Get Sick: the New Science of Darwinian Medicine* [13]. All doctors ought to be familiar with these ideas for they illuminate the nature of illness. It is only within the framework of Darwinian evolutionary theory that one can understand why certain genes have persisted, like those that gave rise to sickle cell anaemia, but protected against malaria and so were adaptive in some countries. It is only within that framework that ageing, cancer, drug resistance and so much else can be properly understood. My own ideas about the biological nature

of depression have their basis in evolutionary theory, namely that depression is an adaptive emotion, sadness, become malignant. Again I have worked hard on the relationship between evolution and embryonic development as it is changes in development due to changes in genes that has resulted in the evolution of animal form and behaviour. We even have an evolutionary model for the origin of multicellularity. While a devoted Darwinian I do puzzle over the gradual nature of evolutionary changes and why the intermediate forms were adaptive.

Neuroeconomics

Presuming that personal choices and decisions are optimally arrived at to benefit survival and reproduction, Neuroeconomics seeks to link patterns of neuronal networks with decision making in humans. This will involve an interdisciplinary approach hailed as "The Second Wave" [14], which Paul W. Glimcher recollects he saw approaching while on the beach,

> Like all academically trained biologists my world view has been shaped both explicitly and implicitly by Darwinian insights ranging from natural selection to the notion that animals have emotions. There is no doubt that Darwin's notion that animals descend from common ancestors shapes almost every facet of what we do as biologists, but it was not until my mid-thirties that I realized, in a profound way, what the rigors of Darwin's natural selection meant for the future of cognitive neuroscience. At that time I was sitting on a beach in Ft. Lauderdale with my collaborator Michael Platt drawing equations and graphs in the sand with sticks. We were there attending a conference on visual perception and had recently seen a number of beautiful talks that described mathematically-based approaches to understanding the function and efficiency of the human visual system. Why, we lamented, were there no similar approaches to the study of higher cognitive behavior in the neurosciences.
>
> At about that time we could not help but notice two seabirds fighting noisily over some morsel of food and it brought to both our minds John Maynard Smith's quantitative parable of the Hawk and the Dove. The theory of games, Maynard Smith pointed out, allows one to assess the efficiency of any competitive behavioral strategy. If the environmental conditions are well defined, choosing to be a hawk a certain percentage of the time and a dove a certain percentage of the time can be unambiguously defined as perfectly efficient.
>
> If natural selection drives generations of animals towards more efficient solutions to behavioral dilemmas, as Darwin proposed, then natural selection must operate upon the neural architecture from which those solutions emerge. What we realized sitting on

that beach was that as long as we could mathematically identify behaviorally efficient strategies then we could identify the direction in which selective pressure must shape the nervous systems of animals. This insight guided us, over the next three or four years, in our search for mathematical definitions of efficient behavioral strategies and ultimately led us to suggest the union of neuroscience and economics in the late 1990s. The result was a highly quantitative approach to the study of decision-making and the nervous system that is coming to be known as Neuroeconomics. More than anything, that venture rests on a quantitative interpretation of the principle of natural selection and an implicit assumption that organisms are driven towards (though not necessarily to) efficient behavior. Absent the law of natural selection, the linkage between models of efficient behavior and neural function on which much of modern neuroscience rests would be impossible.

Evolutionary Psychology

Now we come on to perhaps what has proven the most contentious field to receive the Darwin treatment: evolutionary psychology. Arguments have revolved around such topics as units of selection (genetic determinism), adaptations versus spandrels (panadaptationism), hypothesis testing (falsifiability), interpretative levels (extrapolation) and political versus scientific purpose (motivation). Despite all this antagonism, Steven Pinker has found a 'piece' of mind [15],

> Evolutionary psychology [is] the organizing framework – the source of "explanatory adequacy" or a "theory of the computation" -- that the science of psychology had been missing. Like vision and language, our emotions and cognitive faculties are complex, useful, and nonrandomly organized, which means that they must be a product of the only physical process capable of generating complex, useful, nonrandom organization, namely natural selection. An appeal to evolution was already implicit in the metatheoretical directives of Marr and Chomsky, with their appeal to the function of a mental faculty, and evolutionary psychology simply shows how to apply that logic to the rest of the mind.
>
> Just as important, the appeal to function in evolutionary psychology is itself constrained by an external body of principles -- those of the modern, replicator-centered theory of selection from evolutionary biology-- rather than being made up on the spot. Not just any old goal can count as the function of a system shaped by natural selection, that is, an adaptation. Evolutionary biology rules out, for example, adaptations that work toward the good of the species, the harmony of the ecosystem, beauty for its own sake, benefits to entities other than the replicators that create the adaptations (such as horses

which evolve saddles), functional complexity without reproductive benefit (*e.g.*, an adaptation to compute the digits of pi), and anachronistic adaptations that benefit the organism in a kind of environment other than the one in which it evolved (*e.g.*, an innate ability to read, or an innate concept of "carburetor" or "trombone"). Natural selection also has a positive function in psychological discovery, impelling psychologists to test new hypotheses about the possible functionality of aspects of the mind that previously seemed functionless. For example, the social and moral emotions (sympathy, trust, guilt, anger, gratitude) appear to be adaptations for policing reciprocity in nonzero sum games; an eye for beauty appears to be an adaptation for detecting health and fertility in potential mates. None of this research would be possible if psychologists had satisfied themselves with a naïve notion of function instead of the one licensed by modern biology. [...]

Even if evolutionary psychology had not provided psychology with standards of explanatory adequacy, it has proved its worth by opening up research in areas of human experience that have always been fascinating to reflective people but that had been absent from the psychology curriculum for decades. It is no exaggeration to say that contemporary research on topics like sex, attraction, jealousy, love, food, disgust, status, dominance, friendship, religion, art, fiction, morality, motherhood, fatherhood, sibling rivalry, and cooperation has been opened up and guided by ideas from evolutionary psychology. Even in more traditional topics in psychology, evolutionary psychology is changing the face of theories, making them into better depictions of the real people we encounter in our lives and making the science more consonant with common sense and the wisdom of the ages. Before the advent of evolutionary thinking in psychology, theories of memory and reasoning typically didn't distinguish thoughts about people from thoughts about rocks or houses. Theories of emotion didn't distinguish fear from anger, jealousy, or love. And theories of social relations didn't distinguish among the way people treat family, friends, lovers, enemies, and strangers.

Psycholinguistics

Psycholinguistics falls roughly into two schools of thought about language evolution. Skinner's [16] version of verbal behaviour (now augmented as relational frame theory) proposed learning through Pavlovian conditioned response as brain capacity, intelligence and society gradually co-evolved, in a Darwinian adaptationist fashion. This is in contrast to Chomsky's [17] innate universal grammar. Chomsky proposed a set of linguistic structures hard-wired into our brains, perhaps a functional product of a recursivity spandrel (*sensu* Gould and Lewontin [18]) borrowed from its original problem-solving-in-society context. This reorganisation, and the rise of syntax from it, would not have needed natural selection and so could have been a comparatively rapid process (see Ref. 19 for a recent study), an idea that has not sped by unnoticed: Dawkins included language as part of human culture, and therefore an example of memetics: "Language seems to 'evolve' by non-genetic means, and at a rate which is orders of magnitude faster than genetic evolution" [4].

Now take a moment to listen to the voice in your head. Did you know that the internal dialogue part of our language could be a prerequisite of thought itself [20]? It is especially fascinating to discover that between editions of *The Descent of Man* [21] (the second edition was published in 1882), Darwin became aware of ideas (especially from Ref. 22) that "the use of language implies the power of forming general concepts" and that, "There is no thought without words, as little as there are words without thought". On language evolution he seemed to have a foot in both psycholinguistic camps, suggesting evolutionary development of language as well as instinctual elements [21]: "I cannot doubt that language owes its origin to the imitation and modification of various natural sounds, the voices of other animals, and man's own instinctive cries, aided by signs and gestures [...] may not some unusually wise ape-like animal have imitated the growl of a beast of prey, and thus told his fellow-monkeys the nature of the expected danger? This would have been a first step in the formation of a language [...] The formation of different languages and of distinct species, and the proofs that both have been developed through a gradual process, are curiously parallel". Given that linguistic universals would need to have developed prior to language diversification, Noam Chomsky is understandably non-committal about interpreting language evolution in Darwinian terms,

> I don't separate Darwin from language evolution. He did. He didn't try to address the problem seriously, which is no criticism. [...] The formation of different languages is something that happened AFTER the evolution of the shared human language capacity, hence has essentially nothing to do with evolution of language (apart from what it teaches us about the genetic capacity, shared among humans. As for distinct species, Darwin -- notoriously -- had very little to say about it, apart from some suggestions that have been absorbed into the theory of speciation, which is almost entirely post-Darwin, in fact rather modern. [...]
>
> Darwin has no notion of gradualism in language, apart from a few scattered sentences that could be accepted in almost any approach that is vaguely within the framework of biology [...] The influence of Darwin is that everyone seriously interested in biology (and in my view for the last 50-odd years, language should be studied as part of biology) takes for granted that natural selection is a major factor in evolution. And in some specific areas, that insight has led to very significant achievements. In other areas it has not. Evolution of cognitive capacities is, for the most part, one of these areas. The reasons are pretty clear: the fossil evidence is very slight, and the archaeological evidence thin. So it is necessarily mostly speculation. [I] think there is a possibility of serious work on evolution of language.
>
> Contemporary humans are very similar genetically. They apparently separated about 50,000 years ago, and even apart from interaction, that is far too short a time for any significant evolutionary process to have taken place. Cultural differences are certainly relevant to language research; that's why linguists try to study as many languages as possible. But they are not relevant to research into evolution of language, for the reasons mentioned, except in the indirect sense that language variety tells us a lot about the shared genetic capacity that evolved before the trek from Africa about 50,000 years ago.

While consensus appears to be in favour of some form of pre-adaptation for language capacity, the jury is still in dialogue about subsequent changes that must have taken place, especially those to do with grammatical structure. Ideas revolve around the gradual emergence of a social commentary especially between detached groups and modification of languages via cultural transmission from generation-to-generation. What is clear is the enormous potential for complexity arising from interaction of individual learning, cultural transmission and biological evolution, all adaptive and operating across different timescales, and constrained by learning bottlenecks during childhood [23]. With this in mind, Simon Kirby, a Reader in the Evolution of Language and Cognition at the University of Edinburgh, would push the biological language envelope even further, but within cautionary limits,

> [T]he uniquely human aspect of language may not be as complex as previously thought (see, for example, work by Noam Chomsky and colleagues). If this is the case, then what is special to human may not after all require a complex adaptive explanation. Nevertheless, this view actually highlights those aspects of our language faculty that we may share with other species, thus opening the door to evidence from comparative biology. Whereas linguists have traditionally dismissed the relevance of animal communication to the study of human language, I strongly suspect that future linguistic inquiry will be increasingly informed by what we have learnt from other species. Of this, I am sure Darwin would have approved.
>
> Another recent development in evolutionary thinking that has influenced the study of language has been the popularity of looking at cultural evolution from a Darwinist perspective. One of the most striking features of human language is that it is to a large part culturally transmitted. When a sentence is uttered, it not only conveys semantic information but also information about the particular language of the speaker. Children use this information in combination with their biologically given language faculty to "reverse engineer" the language of the speech community they are born into. When they themselves speak, this process is repeated with the next generation of language learners. There are clear parallels here between the transmission (and therefore evolution) of cultural and genetic information over time. Although we must be careful not to stretch the analogy too far, there have been a number of attempts to show how selectionist mechanisms may apply to the cultural transmission of language. Treating language itself as an evolutionary system may help explain not only how it emerged, but also ongoing language change that is visible today. [...]
>
> So there are several routes along which Darwinian ideas have travelled to converge on the central questions of linguistics, such as why language is the way it is, and how our species came to posses it. When I reflect on the influence of Darwin on my field, however, I realise it is less to do with the specific ways in which we can take an

evolutionary stance on language. In the end these may not be particularly "Darwinian". Rather, I am struck by Darwin's remarkable insight that simple dynamic processes can nevertheless give rise to emergent complexity. In demonstrating the success of this insight, his influence on the study of *all* complex systems cannot be underestimated. Ultimately, he inspires us to take the step from description to explanation.

The Darwinian Resolution
We have briefly explored the evolutionary elements present in evolutionary cosmology, memetics, sexology, Darwinian medicine, neuroeconomics, evolutionary psychology and psycholinguistics. Lewontin would likely welcome the multi-disciplinary nature of these new fields of study. However, along with his fellow critics, he would argue that it is simply not enough for a system to be superficially metaphorically similar to organic evolution by natural selection. If we are so readily prepared to re-apply Darwin's ideas, perhaps we ought to exercise the same standard of scientific rigour with which he developed and applied them. But what would Darwin think?

All opinions are those of the author unless clearly indicated otherwise. Unless the literary source is otherwise indicated, all quotes were obtained through direct communication with each named contributor, and are used with permission for this article. Thank you to all of the contributors.

John Bancroft is Director Emeritus of The Kinsey Institute for Research in Sex, Gender and Reproduction at Indiana University. He has been involved in various aspects of sex research for the past thirty years and is a world authority on the relationship between reproductive hormones and sexual behavior, and has extensive research experience in fertility control and its relevance to sexual behavior, psychophysiological aspects of male sexual response, and the impact of the menstrual cycle on the sexuality and well-being of women. He has published more than 200 papers related to sex research and is the author of *Human Sexuality and Its Problems* (1989).

Susan Blackmore is a freelance writer, lecturer and broadcaster, and a Visiting Lecturer at the University of the West of England, Bristol. Her research interests include memes, evolutionary theory, consciousness, and meditation. Her books include *The Meme Machine* (1999) and *Consciousness: An Introduction* (2003).

Noam Chomsky is an Institute Professor emeritus and Professor Emeritus of Linguistics at the Massachusetts Institute of Technology. Since beginning his theory of generative grammar, his works have had a major influence on theoretical linguistics and cognitive science. He is a prominent critic of US foreign and domestic policy, and has been cited more often than any other living scholar between 1980–1992, and was ranked the eighth most-cited ever. He has written about thirty books on linguistics, one on computer science and over seventy political works, most recently, *Interventions* (2007) and *What We Say Goes: Conversations on U.S. Power in a Changing World* (2007).

Paul W. Glimcher is Professor of Neural Science, Economics and Psychology at New York University. He has authored *Decisions, Uncertainty and the Brain: The Science of Neuroeconomics* (2003) and has co-edited *Neuroeconomics: Decision Making and the Brain* (with Colin Camerer, Ernst Fehr and Russell Poldrack, 2008).

Simon Kirby is Reader in the Evolution of Language and Cognition at the University of Edinburgh. He has written many academic articles and invited chapters, plus two well-received books, *Function, Selection and Innateness: the Emergence of Language Universals* (1999) and *Language Evolution* (with Morten Christiansen, 2003).

Richard C. Lewontin is Alexander Agassiz Professor of Zoology and Professor of Biology at Harvard University. His work with John L. Hubby in 1966 helped establish the modern field of molecular evolution, and he introduced "spandrels" into evolutionary biology with Stephen J. Gould. His most recent book is *Biology Under the Influence: Dialectical Essays on Ecology, Agriculture, and Health* (with Richard Levins, 2007).

Steven Pinker is a Harvard College Professor and the Johnstone Professor of Psychology, known for championing the idea that language is an instinct or biological adaptation shaped by natural selection, in opposition to the Chomskyan model. His experimental research on cognition and language won the Troland Award from the National Academy of Sciences, the Henry Dale Prize from the Royal Institute of Great Britan, and two prizes from the American Psychological Association. He has also received several honorary doctorates and numerous awards for graduate and undergraduate teaching, general achievement, and his books which include, *The Blank Slate: The Modern Denial of Human Nature* (2002) and *The Stuff of Thought: Language as a Window into Human Nature* (2007). Stephen submitted his previously published foreword to David Buss's *Handbook of Evolutionary Psychology* [16].

Rupert Sheldrake is the current Perrott-Warrick Scholar and Director of the Perrott-Warrick Project. He is also a Fellow of the Institute of Noetic Sciences, near San Francisco, and an Academic Director and Visiting Professor at the Graduate Institute in Connecticut. His extensive writings include *The Presence of the Past: Morphic Resonance & the Habits of Nature* (1988) and *Dogs That Know When Their Owners Are Coming Home: And Other Unexplained Powers of Animals: An Investigation* (1999).

Lewis Wolpert is Professor of Biology as Applied to Medicine in the Department of Anatomy and Developmental Biology of University College, London. He was made a Fellow of the Royal Society in 1980, a Fellow of the Royal Society of Literature in 1999 and awarded the CBE in 1990. He is also a Vice-President of the British Humanist Association. His books include *Malignant Sadness: The Anatomy of Depression* (1999) and *Six Impossible Things Before Breakfast: The Evolutionary Origins of Belief* (2006).

1. Darwin, C.R. (1859) *On the Origin of Species by Means of Natural Selection, or the Preservation of Favoured Races in the Struggle for Life*. John Murray, London.

2. Smolin, L. (1992) Did the Universe Evolve? *Classical and Quantum Gravity* 9, 173-191.

3. Darwin, C.R. (1872) *The Expression of the Emotions in Man and Animals*. John Murray, London.

4. Dawkins, R. (1976) *The Selfish Gene*. Oxford University Press.

5. Blackmore, S.J. (1999) *The Meme Machine*. Oxford, Oxford University Press.

6. Benítez-Bribiesca, L. (2001) Memetics: A dangerous idea. *Interciencia* 26, 29–31.

7. Schrödinger, E. (1944) *What Is Life?* Cambridge University Press (1992 imprint).

8. Herzog, D. (2006) The Reception of the Kinsey Reports in Europe. *Sexuality & Culture* 10, 39-48.

9. Burkhart, R.A. (1948) Church Can Answer the Kinsey Report. *Christian Century* 65, 942-943.

10. Kinsey Report Criticized: Dr. Bonnell Cites Defects, but Says Church Can't Ignore It. *New York Times* Oct. 26, 1953. Available online http://www.nytimes.com/ads/kinsey/akinsey_5.html [Accessed 14/10/08]

11. Kinsey, A.C., Pomeroy, W.B., Martin, C.E. and Gebhard, P.H. (1953) *Sexual Behavior in the Human Female*. W.B. Saunders, Philadelphia, PA.

12. Williams, G.C. and Nesse, R.M. (1991) The dawn of Darwinian medicine. *Quarterly Review of Biology* 66, 1-22.

13. Nesse, R.M. and Williams, G.C. (1994) *Why We Get Sick?: The New Science of Darwinian Medicine*. Times Books, Random House, NY.

14. Hammerstein, P. and Hagen, E.H. (2005) The Second Wave of Evolutionary Economics in Biology. *Trends in Ecology and Evolution* 20, 604-609.

15. Buss, D.M. (ed.) (2005) *Handbook of Evolutionary Psychology*. Chichester: John Wiley & Sons.

16. Skinner, B.F. (1957) *Verbal Behavior*. Copley Publishing Group, Acton, Massachusetts.

17. Chomsky, N. (1965) *Aspects of the Theory of Syntax*. MIT Press.

18. Gould, S.J. and Lewontin, R.C. (1979) The Spandrels of San Marco and the Panglossian Paradigm: A Critique of the Adaptationist Programme. *Proceedings of the Royal Society of London* B 205, 581-598.

19. Atkinson, Q.D., Meade, A., Venditti, C., Greenhill, S.J. and Pagel, M. (2008) Languages evolve in punctuational bursts. *Science* 319, 588.

20. Chomsky, N. (2005) *Some Simple Evo-Devo Theses: How True Might They Be For Language?* Alice V. and David H. Morris Symposium on Language and Communication; The Evolution of Language. Stony Brook University, New York, USA (October 14 2005). Available online http://www.punksinscience.org/kleanthes/courses/MATERIALS/Chomsky_Evo-Devo.doc [Accessed 14/10/08].

21. Darwin, C.R. (1871) *The Descent of Man, and Selection in Relation to Sex*. John Murray, London.

22. Müller, F.M. (1873) Lectures on Mr. Darwin's Philosophy of Language. *Fraser's Magazine* 7, 525–41, 659–78, 8, 1–24. In R. Harris (ed.), *The Origin of Language*, pp. 147–233. Thoemmes Press, Bristol, 1996.

23. Christiansen, M.H. and Kirby, S. (2003) Language Evolution: Consensus and Controversies. *Trends in Cognitive Sciences* 7, 300-307.

Water location, piospheres and a review of evolution in African ruminants

Introduction

This paper sets out to revise our understanding of grazer evolution in Africa. In it we briefly review the current model before submitting for your consideration additional mechanisms that may have lead to speciation and diversification. In doing so we extend the concept of species refugia and rebalance our understanding of processes with respect to the priorities facing the survival of populations under a drying climate.

Firstly, we know that the main radiation of large mammalian herbivores in Africa took place during the Pliocene-Pleistocene (Sinclair 1983), between 5.3 million years ago (mya) and 10 thousand years ago (kya). For example, recently discovered fossil-evidence from the Turkana Basin of Kenya and Ethiopia indicate 58 to 77% turnover of the mammal species between 3.0 and 1.8 mya (Behrensmeyer *et al.* 2003) where species diversity increased from 3.0 to 2.0 mya, but declined thereafter.

Habitat change in the Pliocene-Pleistocene

Additionally, a range of evidence from deep-sea cores in the Atlantic offshore, sedimentology of pluvial lake sediments, and the available palynological work indicate that it was after 3.0 mya that a higher level of aridity became established generally over Africa (Street-Perrott and Gasse 1981, Goudie 1996) and specifically across southern Africa (Thomas and Shaw 1991, 2002). Distinct periods of increased aridity occurred at 2.8, 1.7 and 1.0 mya (deMenocal and Bloemendal 1995). These epochs were marked throughout by a fluctuating climate; a continuous oscillation with an initial periodicity of 19,000-23,000 years prior to 2.8 mya, increasing to 41,000 years after 2.8 mya, and 100,000 years after 1.0 mya (deMenocal and Bloemendal 1995); with a long-term trend towards increased aridity (Adams 2000, deMenocal 2004, Gasse 2006).

Rifting caused changes to precipitation patterns in Africa, particularly eastern Africa, from as early as the late Miocene (*e.g.*, WoldeGabriel *et al.* 2001). This effected tree cover and distribution, because there is a minimum threshold of rainfall which is required to maintain the closed-canopy of woodland (Sankaran *et al.* 2005), and topography can give rise to localised rainshadow effects (Sepulchre *et al.* 2006). So, aridification during the Pliocene-Pleistocene very likely led to a complex response in environmental conditions: notably, the drier climate promoted the expansion of grasslands in Africa at different times and at different localities, with the potential for neighbouring regions experiencing identical climate change to produce contrastingly different habitats: for example, fossil deposits from the west Turkana region and the lower Omo valley of Ethiopia indicate open vegetation close to riverine forest (Bobe 2006). The variable climate caused the environment to repeatedly cycle between the previously dominant forest and newly invasive open grasslands; savanna, the intermediate habitat type, would have been present twice as long as any other habitat type (Sinclair 1983).

African surface water distributions are also rain fed and, therefore, they are dynamic. Well known examples include Lake Chad, which 6 kya covered an area of 400,000 km^2. It has dried up several times, about every 3,000 years, and has decreased from 26,000 km^2 to 1,500 km^2 in the last 50 years (UNEP 2002). During the greater aridity of the Pliocene-Pleistocene, many of the lake systems shrank significantly and periods of dune formation are seen (Goudie 1996). While precise dating of historical periods of aridity remains restricted to the last 50,000 years (Thomas and Shaw 2002), there is clear evidence of arid phases being associated with the isolation of river and lake systems: for example, from anomalies in the species assemblages of fossil fish and crocodile assemblages (Grove *et al.* 1975, Beadle 1981). Seismic reflection imaging and core analysis show that Lake Victoria was dry for about 5,000 years during the late Pleistocene (Johnson *et al.* 1996), while certain components in other fossil assemblages suggest that between 135 and 70 kya, Lake Malawi was brackish and nearly 600 m shallower than at present (Cohen *et al.* 2007).

The isolation of surface water systems plus increased dune activity (Lancaster 1989) strongly suggests that greater distances would have existed between sources of water available for animal consumption. But, our current understanding of animal evolution during this period focuses on habitat change and only goes some way towards incorporating this fragmentation of surface water systems.

Herbivore evolution in the Pliocene-Pleistocene

The expansion of grasslands would have been assisted by the consumptive impacts of forest-dwelling herbivores (Bobe 2006) plus tree removal by fire (Beerling and Osborne 2006). These spreading grassland habitats and grassland-woodland mosaics favoured the diversification of ruminant grazers in Africa, in contrast to the earlier rise of hind-gut fermenters in North America, 18 mya in the mid Miocene (Janis *et al.* 2002a). About this time, falling sea levels had allowed grazers to migrate into Africa from Eurasia (Janis 2008), having originated in Asia, 40 mya during the Oligocene (Prins 1998). The traditional explanation of subsequent bovid evolution in Africa during the Pliocene-Pleistocene posits an "adaptive radiation within browser and grazer types along food quality gradients" (McNaughton and Georgiadis 1986) and so relates animal adaptations to the habitat modification brought about by climatic change (*e.g.*, Janis 1989, Bobe and Eck 2001, Behrensmeyer *et al.* 2003, Barnosky 2005).

Vrba's *resource-use hypothesis* (Vrba 1980, 1987, Fernández and Vrba 2005) largely contributed to this model by identifying the selection pressures that are associated with these physical changes to the habitat as the principal factors in vicariance, the shrinkage or fragmentation of a species' geographic distribution that leads to speciation. Vrba termed environmental change as 'habitat drift', and a species' response to it as 'distribution drift'. Habitat drift is caused by changes in the resources that an animal relies on for survival: "temperature, moisture, substrate, diverse physical and chemical constituents, other organisms, food items, vegetation cover and any other environmental components", including water to drink. A prediction of her hypothesis was that resource generalists are more able to survive such changes than specialists (see Vrba 1992 for a synopsis of her *habitat theory*).

Of course, the fossil record attests for the huge numbers of animals that were unable to survive events during the Pliocene-Pleistocene. Various explanations have been offered to account for the global mass extinctions in the bovidae (*e.g.*, Vrba 1980, Behrensmeyer *et al.* 2003), particularly in the late Pleistocene (*e.g.*, Owen-Smith 1987). Habitat change through climate change is the most enduring explanation, but the coincidental increase in human hunting pressure was also thought to have largely contributed (Martin and Klein 1984). Owen-Smith's *keystone-herbivore hypothesis* (Owen-Smith 1987, 1989, Zimov *et al.* 1995) proposed that populations of medium-sized grazers could have tracked suitable habitat conditions by following corridors of open, nutrient-rich vegetation created by megaherbivores (Owen-Smith 1988), until mainly human hunting removed those megaherbivores. With their demise so ended the facilitation. Alternatively, humans were the cause of mass extinctions as the source of disease according to the *hyperdisease hypothesis* (MacPhee and Marx 1998). There is evidence in support of each of these hypotheses, but exactly what caused the megafaunal extinctions remains unknown, although it is likely that there were different reasons on each continent (Barnosky *et al.* 2004); Guthrie (2006) recently disputed any direct human influence on the Pleistocene extinctions in North America, suggesting that we had had a more indirect impact on animal survival.

Adaptation to habitat change with the rise of water dependency
It is assumed that the small-bodied ancestral browser in Africa was relatively water independent, as are typical contemporary browsers, gaining most of their physiological water from their diet. But when the development of open, dry, savanna grasslands started early in the Pleistocene (Janis 2008), competing, browsing herbivore populations would have been presented with a strong selective pressure for conversion to grazing; episodic habitat changes could have produced grazers separately on more than one occasion (Essop *et al.* 1997, Janis 2008, but also Georgiadis *et al.* 1990), while diversification would have been accelerated by a switch from monogamy to polygyny (Pérez-Barbería *et al.* 2002).

Adaptation to grazing involved an increased efficiency in digesting fibre, an ability that likely co-evolved with the fibre content of plants, and the rate of which followed a species-specific adaptation model that shows a gradual acceleration over time, possibly in response to increased competition between emerging, grazing species (Pérez-Barbería *et al.* 2004). Gordon and Illius (1996) have shown that this change in feeding strategy necessitated an increase in body size, and this increase is indeed confirmed by the fossil record (Janis 1989), although, intake limitation by the foregut means that this increase was not without limit (Clauss *et al.* 2003). Interestingly, if trophic competition is limited or even entirely absent, as found for some island species, the tendency is for "the Island Rule" to reduce body size (Raia and Meiri 2006). This even occurs on islands where predators are present, which also suggests that diet is the principal driver of body size evolution (see Hoffman 1989 for a contradictory viewpoint that favours anatomical adaptations to diet quality independent of body size, in contrast to body size determination by diet type because of metabolic requirements, as asserted by Gordon and Illius 1996, Gagnon and Chew 2000, *etc.*). Nonetheless, even slight increases in body mass could have been advantageous for reasons other than diet: particularly when fighting over mating access to females (Pérez-Barbería *et al.* 2002), defending against predators and the benefits of possessing a longer gait (Hudson 1985). So, while small animals avoid some of the heightened predation near water holes by being less water dependent (Ayeni 1975), larger species are better able to 'fight' or 'flee' (Scott 1985).

However, and most importantly, with their adaptation to poorer and drier herbaceous diets, animals incurred a cost: an increase in water requirements to aid digestion and to thermoregulate their larger body size[1]. Vrba's analysis (Vrba 1987) showed that stenotopic water dependence (*i.e.*, able to tolerate only small changes) is so critical that it nullified any survival benefits of having evolved a non-specialist diet. Therefore, to account for how water-dependent populations did survive, Vrba (1987) proposed a concept of wet refugia as places where the water-dependent impalas (*Aepyceros melampus*) and African buffaloes (*Syncerus caffer*) may have survived habitat drift and

---

[1] Animals need to maintain a water balance within acceptable limits for homeostatic function. Lethal levels of water loss are reported for mammals at 20-36% of body weight (Adolph 1943 cited in Peters 1983).

the depletion of their drinking water. Notably, wet refugia were suggested as ways in which these grazing genera may have persisted, relatively unchanged; they exhibit unbranching lineages, implying that their isolation in wet refugia did not produce speciation, whereas others (Kingdon 2003, Bobe 2006) have proposed that major river valleys could have acted in a similar way, but as corridors for forest and woodland mammals, and centres of endemism for those browsing species, somewhat like gorillas in Central Africa (Anthony *et al.* 2007). These low speciation rates are especially surprising considering the evidence for rapid evolution by small isolated populations (Vrba and DeGusta 2004) and divergence through genetic drift once isolated (Knowles and Richards 2005). However, Vrba (1987) explained the incongruity by suggesting that because water sources are so critical, removing them produces a local extinction rather than leading to speciation, so very few isolated populations survived long enough to register diversification in the fossil record.

At this juncture, note how animal water requirements are included in the current model of bovid evolution in Africa; water dependency relates a physiological need to an environmental resource and this is more critical in its requirement than other needs; reproductive isolation occurred via habitat fragmentation as a consequence of climate change; habitat fragments would have contained sources of drinking water for the persistence of water-dependent species; the *location* of drinking water and its influences are subsumed within the extent of those habitat fragments.

The scale of vicariance

The current model of bovid evolution in Africa operates at a spatial scale defined by habitat extent: populations become isolated because of the distances that separated habitat fragments. And yet it is evident that water dependency means that the opportunity for an animal to forage is defined by the interval between drinking events (*e.g.*, Tolkamp *et al.* 1999). Thus, the need to forage is constrained by the requirement to drink, and ranging patterns (determined by foraging behaviour) are constrained by the location of water, to differing degrees as dictated by animal water dependency (Ayeni 1975). This means that while browser distributions are relatively unaffected by the location of drinking water (Redfern *et al.* 2003), under our present climate, grazing ranges are most constrained during the dry season (Ayeni 1975), when the combination of digestive constraints and travel costs transform free-ranging, water-dependent animals into central-place foragers, thereby limiting their capacity to utilize the full extent and potential of their wet-season habitat (Derry 2004): even year-round, and particularly for the most water-dependent species, "waterholes create nodes of attraction within suitable habitat" (Fig. 6 in Smit *et al.* 2007). Under the Pliocene-Pleistocene climate each 'dry season' would have lasted for several thousand generations: a sufficient period for random drift to fix six neutral mutations in static, large herd populations of buffalo of at least 1,600 individuals, and twenty-eight in equivalent, average-sized herds of 350 individuals (van Hooft *et al.* 2003), [based on Kimura and Ohta's (1969) estimate of time to fixation (but also see Kumar and Subramanian 2002, and Gillooly *et al.* 2005), and using an estimated 5 year generation (doubling) time for buffalo (Van Sickle *et al.* 1987)].

In summary, the current model identifies habitat fragmentation in response to climate change as the primary cause of speciation and diversification. However, it is less clear what role the location of drinking water had to play in this habitat fragmentation and the subsequent evolutionary processes. The current model neglects to consider the interaction of water dependency with the location of drinking water; increased aridification would not only have involved shifts in the spatial distribution and species composition of forage resources, but it would also have increased the patchiness of surface water distributions, causing animal foraging distributions to be constrained within specific distances of their drinking water over an evolutionary timescale; distances which were not necessarily in accord with the extent of their potential habitat as solely defined by forage distributions, particularly as animals became more dependent on water with the increase in grazing and its associated traits. In terms of the resource use hypothesis, water dependency under arid conditions would have given rise to discord in the otherwise potentially synchronous match between habitat drift and distribution drift.

The hypothesis presented here proposes that the consistently adopted metric "distance to water" (*e.g.*, Heshmatti *et al.* 2002, Redfern *et al.* 2003, Nangula and Oba 2004, Bisigato *et al.* 2005, Smet and Ward 2006, Chamaillé-Jammes *et al.* 2007) is the appropriate operational scale for distribution drift under conditions when water is limiting. In doing so it extends Vrba's concept of wet refugia (Vrba 1987) by suggesting that a more heterogeneous distribution of surface water resulted from a drying climate that had the potential to introduce spatial structure into a comparatively homogeneous environment, and thus introduce constraints on animal movements and ultimately on gene flow. This paper initially tests the influence of water dependency on habitat selection, and then, satisfied of its importance, outlines the potential for the location of drinking water to play a role in the evolution and diversification of these African large mammalian herbivores and proposes possible ways in which to test this novel hypothesis: regarding the obvious gaps in existing evidence, it was once noted that, "[s]keptics tend to preach that the interesting questions are unanswerable, that promising approaches are in reality flawed by insuperable methodological shortcomings and hopelessly inadequate data, and that anything that might have been worth saying was said long ago" (Vermeij 1989). However, in the apparent absence of any scientific literature attempting to approach this topic, this paper is intended to stimulate debate; it simply suggests the potential roles of water location in influencing animal evolution, while recognising the need for further work to establish the likelihood of these mechanisms.

For the purposes of this paper, a 'watering point' is used to denote a single location from which animals may access a source of drinking water, say, on the banks of a pan. This is equally applicable for larger water bodies, dambos and riverine systems; animals are known to habitually access rivers at certain locations where the river bank may be shallow enough, or drainage lines from the surrounding landscape have eroded steep banks, thereby providing easier access to the water (Thrash and Derry 1999). 'Water sources' are assumed to be water bodies providing one, or more, watering point, or points.

The role of water in African large mammalian herbivore evolution

The influence of water dependency

Simply put, animals die without water. Water must be an integral part of any animal's habitat (Aristotle 350BC), so it is reasonable to expect that it must have been in sufficient supply to support populations who became isolated following the fragmentation of their habitats. The question remains whether, the location of drinking water was purely a minor influence on long-term habitat selection, or the location of drinking water actually defined the extent of their environment.

For water location to be a secondary factor in determining habitat selection, either animals do not include it in their assessment of habitat viability along with other factors, such as forage composition (*e.g.*, Fritz *et al.* 1996) and shade (*e.g.*, van Heezik *et al.* 2003), or drinking water itself is sufficiently abundant so as not to limit animal foraging behaviour. The latter was the situation found by Prins (1996) who concluded that there was enough water in Lake Manyara National Park that it did not limit the choice of grazing grounds by buffalo. But water is not always in such copious supply, and being indifferent to where you are going to find drinking water, and assuming that it is always associated with the habitats that you choose based on other criteria, would be a surprising evolutionarily stable strategy.

Unsurprisingly therefore, a spate of recent studies have found water location to be a principal concern, if not the most important consideration, involved in habitat selection (*e.g.*, Traill 2004, Roux 2006, Henly *et al.* 2007, Reid *et al.* 2007, Smit *et al.* 2007). And when water is less ubiquitous, for example, during severe droughts, water-dependent animals such as buffalo remain near the enduring water sources (Ryan and Jordaan 2005). Henly *et al.* (2007) concluded that "under conditions of extreme aridity [...] an area is selected primarily for the presence of a reliable water supply, and thereafter habitat selection occurs at a smaller spatial scale within that area on the basis of the plant parameters". This strongly suggests that animals do "know" the value of "knowing" where drinking water is to be found. The alternative would be too costly; drought-related mass mortalities of livestock and wildlife when drinking water does become limited are all too apparent (*e.g.*, Sinclair and Fryxell 1985, Walker *et al.* 1987, Knight 1995).

The influence of water location

Ideal Free Theory (Fretwell and Lucas 1970) predicts that herbivore densities should reflect resource distributions in heterogeneous environments, assuming that animals have perfect (ideal) knowledge of resource profitability and are free to move between resource sites. The effect is that animal impacts are distributed across landscapes according to the spatial distribution of their forage (Tyler and Hargrove 1997), as an Ideal Free Distribution (IFD). Water-dependence forces a deviation from the IFD through restriction of animal foraging, thereby compromising animal foraging efficiency as a function of resource supply (predominantly food quality in grazers, *e.g.*, Wilmshurst *et al.* 1999 and Shrader *et al.* 2006).

The local accumulation of impacts associated with the congregation of animals at water points and the declining grazing pressure with distance from water gives rise to a utilisation gradient termed the "piosphere" pattern (Lange 1969 plus see review by Thrash and Derry 1999). Available foraging area decreases rapidly on nearing the watering point (Perkins 1991) and, as animals move towards it, their cumulative foraging effort becomes concentrated into less space. The result is a gradient of grazing intensity (Andrew 1988), which is greatest nearest the watering point and decreases as a function of distance from it, until reaching the furthest distance from water an animal may travel during the period before returning to drink. This defines the extent of the piosphere (Graetz and Ludwig 1978), a maximal model for available foraging area at distance from water. Real world examples are more likely contained by landscape features that influence animal movement, and may not conform to geometric prediction (*e.g.*, Weir 1971), largely because of variations in the animals' behavioural response to wind, topography and spatially heterogeneous vegetation (Nash *et al.* 1999).

Piospheres are most pronounced in arid and semi-arid zones, where animals are most dependent on drinking-water sources. Examples of African piosphere patterns have been reported in herbaceous (Abule *et al.* 2005) and woody (Perkins 1991) species compositions, range condition (Getzin 2005), grass production (Edroma 1989), biomass and defoliation (Mphinyane 2001, Derry 2004), basal (Nangula and Oba 2004) and canopy (Stokes and Yeaton 1994) covers, soil quality (*i.e.*, soil pH, nitrogen, and organic carbon) (Smet and Ward 2006) and other plant nutrients (Moleele 1994), rainfall infiltration (Thrash 1997) and remotely sensed vegetation indices (Washington-Allen *et al.* 2004). In areas populated by elephants, tree density and cover have also been found directly proportional to the distance from a water point (Brits *et al.* 2002).

The extent of the impact, as reflected by the response of a measured variable, and our interpretation of that response, depends upon which variable is being measured (Fernandez-Gimenez and Allen-Diaz 1999), when it is being measured (in terms of season and the age of the site) and where it is being measured (*i.e.*, dependencies on climate and vegetation type). For example, a gradient in soil surface characteristics may extend to only a few tens of metres away from a watering trough (Andrew and Lange 1986), whereas trends in herbaceous plant basal cover may be detected up to 7 km from the focal point (Thrash *et al.* 1991). This is the reason why water dependence is considered most detrimental to animals during the dry season, because it restricts their foraging range to these areas that have already been degraded by high utilisation pressures (*e.g.*, Ryan and Getz 2005) and this ultimately limits their food intake rate (Derry 2004).

The influence of water distribution
Areas that offer forage but lack drinking water are inaccessible for utilization by water-dependent animals (Owen-Smith 1996). That is why management policies in game reserves and other protected areas within African semi-arid regions have typically introduced water sources to "open up" areas (du Toit and Cummings 1999), historically advocating an even distribution of artificial water sources to increase the carrying capacity of the land by accessing waterless zones and distributing the grazing pressure evenly. Notorious examples include the Kalahari Gemsbok National Park (Perkins 1996) and the Northern Plains of Kruger National Park (Pienaar *et al.* 1997, Braack

1997). But during the Pliocene-Pleistocene, with no human intervention and under a drying climate, a patchy natural distribution of water sources would have resulted if comparatively ephemeral water sources had dried out as the climate became more arid.

Evaporation is related to surface area: water loss from a circular water source increases exponentially with its diameter (Zambatis 1985). Insufficiently sized water sources (<10 m diameter) tend to become mud wallows. This deters drinking. Larger water sources (>25 m diameter) lose about 920 m^3 of water per annum (Zambatis 1985). Chamaille-Jammes *et al.* (2007) showed that it required rainfall levels in excess of 900 mm per annum before non-riverine water sources were more likely to hold water than riverine systems. Under the most extreme dry conditions that they measured (200 mm per annum), there was only a 10% chance of finding static surface water, clearly emphasising animal reliance on riverine systems. Under predictions of current climate change, a 10% reduction in southern African rainfall is forecast to result in 23% less surface-water flow through perennial rivers (de Wit and Stankiewicz 2006); 20% less rain would halve river flow (Fig. 1). Unfortunately, there are few estimates of African precipitation levels during the Pliocene-Pleistocene; rare examples were given by Cohen *et al.* (2007) who estimated <400 mm per annum for the Lake Malawi watershed, and Bonnefille (1984, cited in Egeland 2007) who estimated the low precipitation levels at Olduvai Gorge in northern Tanzania to have been about 350 mm per annum between 1.76–1.75 mya, suggesting limited perennial river flow for 10,000 years.

As the water sources dried out, gaps between the habitat fragments which contained water sources would have expanded, making it impossible for animals to migrate between those fragments. A modern day example can be seen in Kruger National Park, which receives water from five perennial rivers that traverse the park (Rogers and O'Keefe 2003). These rivers can be considered the only long-term permanent water sources, as other rivers and pans within the park are ephemeral or depend on the accumulation of rainfall, or are artificially provided (Gaylard *et al.* 2003). Without these temporary water sources, animal movement between the perennial rivers would not be possible: they are separated by distances of 35 km to 180 km (Gaylard *et al.* 2003), at least twice the distance that extant, water-dependent, large herbivores are typically found from their water sources during the dry season, *e.g.*, 10-15 km in Amboseli (Western 1975) and 12-16 km in Kruger National Park (Van der Schijff 1957). Certainly, during the dry seasons before the artificial water sources were introduced, animals would indeed be found concentrated along the rivers (Gaylard *et al.* 2003) as "most of the water sources in the areas between the rivers were unreliable" (Smit *et al.* 2007); Ayeni (1977) also found that pans in Tsavo National Park would dry out within three months of the rains ending. A larger, areal example ($\sim 5 \times 10^6$ km^2) can be found in Bauer *et al.* (1994), who clearly showed animals and rivers in proximity in north-east Africa.

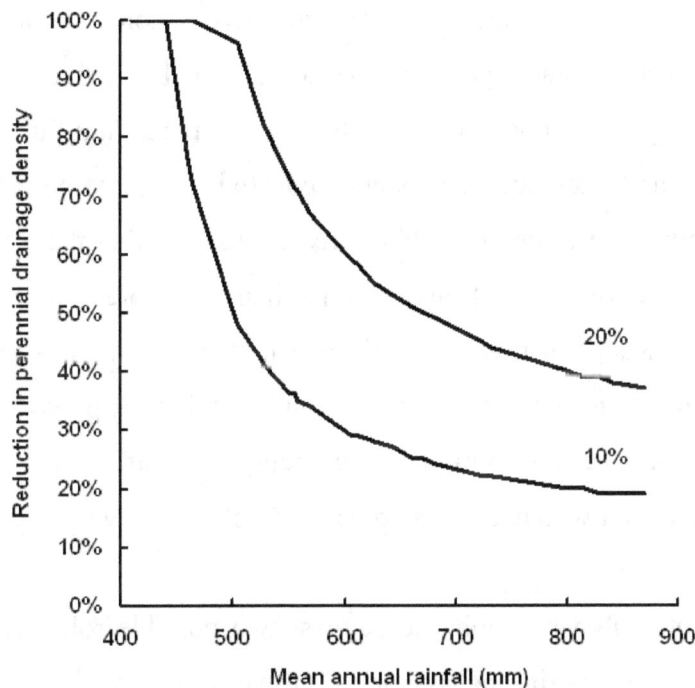

Fig. 1: Percentage reduction in drainage density (the length of stream channel per unit area of drainage basin, and therefore an estimate of runoff potential) for perennial rivers in Africa, as a function of mean annual rainfall, under two scenarios: A) a 10% reduction in mean annual rainfall, and B) a 20% reduction in mean annual rainfall. Data from Table 1 in de Wit and Stankiewicz (2006).

Imagine an area divided into various habitats that contain a range of densities of water sources, from homogeneous through to waterless. An one extreme there would be sparse distributions resulting from a few evenly spaced water sources. At the other, fewer clusters of multiple water sources. Common to each distribution would be the potential to distance populations and prohibit mixing.

The influence of watering behaviour

Critically, for evolution, the question remains whether isolation of animal populations during periods of aridification occurred purely as as a consequence of the fragmentation of food distributions, and to what degree it involved increasingly heterogeneous distributions of water sources. Unfortunately, ancient water and forage distributions are not available to answer these questions, nor are there records of animal water dependency and its effect on foraging distance. However, it is likely that water location limited the foraging ranges of extinct animals in the same way that it influences modern animals (Derry 2004), and that water dependency constrained the home range of free-ranging animals within more extensive habitat areas, thus dictating the availability of their food. We also know that extinct large mammalian herbivores possessed similar femoral morphologies (Kappelman *et al.* 1997) and moved at similar speeds to their modern day descendants (*e.g.*, Janis *et al.* 2002b), and that they employed some of the strategies seen in extant species to minimize the energy costs of locomotion, *e.g.*, by adopting a slow walking gait with reduced step length (Köhler and Moyà-Solà 2001).

The influences presented above combine to suggest two possible roles, acting across two spatial scales, for the watering behaviour of African ruminants in bringing about morphological adaptations and driving the diversification of their feeding strategies:

(1) *inter-patch distance from water* - as patches of persistent drinking sources became isolated over evolutionary time, the distance between these wet refugia may be considered a geographical barrier akin to any other barrier to animal migration and the mixing of populations.

(2) *intra-patch distance from water* - selection pressures act within patches of drinking sources in response to the concentration of herbivore densities about watering points.

Evolutionary mechanisms

This section discusses the conditions under which the ecological consequences of water dependency may have played a role in animal diversification and, particularly, the dietary adaptations of grazing. In each case, a spatial distribution of water sources implies an evolutionary mode of speciation: single patches for sympatric speciation, and multiple patches for allopatric and parapatric speciation. There are also suggested mechanisms for the reinforcement of introduced differences between divergent populations. For a detailed discussion about these theories see Coyne and Orr (2004).

i. <u>Sympatric speciation</u>

Water dependence restricts an animal's foraging range during the dry season to areas degraded by high utilisation pressures. High water dependency is associated with adaptations to a grazing diet (*e.g.*, large body size). It is therefore reasonable to expect a piosphere to produce opposing selection pressures: to get further away from water sources, or adapt to the consequences of the utilisation gradient and the associated limitation of the forage supply and vegetation compositional changes (Pennycuick 1979).

Limitation of forage along the utilisation gradient is manifested in the herbivore functional response (Derry 2004), while the plant response may lead to changes in species composition, depending on the life histories of individual plant species present and on their location along the gradient of grazing pressure (Lailhacar *et al.* 1993). Grazers and grasses have co-evolved (Stebbins 1981): grasses became more silica-rich and fibrous, and less palatable in response to being eaten, and herbivore teeth adapted to the increasingly abrasive diet (Mendoza and Palmqvist 2008). Furthermore, some studies have identified a trend for displacement of higher quality grasses towards the perimeter of the piosphere (*e.g.*, Friedel 1988, Perkins and Thomas 1993, Thrash *et al.*

1993). Ergo, piosphere generation and maintenance may be considered an example of how animals can create or destroy their own ecological niches with evolutionary consequences, a process called "niche construction" (Odling-Smee *et al.* 2003), though synonymous with ecological engineering (Jones *et al.* 1994). As grazing developed, so animals would have frequented water sources more often, accumulating their impacts and causing a reduction in forage quality and abundance: piosphere development would likely modify the selection pressures acting directly on the animal engineers. This is supported by the gradual increase over evolutionary time of fibre in bovid diets (Pérez-Barbería *et al.* 2004). In addition to protracted competition between grazers (Pérez-Barbería *et al.* 2004), the longer climate periodicity would have meant longer periods of isolation between water sources, extending the periods of niche construction and plant-animal co-evolution. Piosphere plant responses may have driven that adaptation to poorer quality diets.

This notion is supported by field observations of overlap in diet between grazing herbivore species. The overlap is highest during the wet season, but declines from the start of the dry season as each species concentrates on niches of "food refuge" (Sinclair 1983, du Toit *et al.* 1995, Traill 2004 but also see Prins and Fritz 2008 for a wider discussion). In the late dry season the overlap increases again as resources are depleted to an extent that forces compromises in diet selection, suggesting that interspecific competition promotes ecological separation by food preference. Therefore, during the dry season, adaptations promoting niche separation are most strongly selected (Schoener 1974)[2].

---

[2] Illius and Gordon (1991) predicted that maximum daily energy intake should scale with body size as $W^{0.88}$ (*i.e.*, greater than metabolic rate, $W^{0.75}$), implying an advantage of large size for utilisation of diets with slow digestion rates, *i.e.*, larger species can subsist on poorer-quality food. The result is a selective pressure for adaptation to a poorer quality diet, as is typical for depleted dry-season resources (Codron *et al.* 2007b), which carries with it a corresponding increase in body size (Illius and Gordon 1992). It also suggests a divergence of body size via a strong selection pressure for large body size during periods when forage is abundant but of poor quality, countered by the pressure for small size during seasons when forage is either abundant (Illius and Gordon 1992), or scarce, but of high quality (Murray and Illius 1996). Codron *et al.* (2007a) recently suggested that physical differences between diets may also account for slower digestion rates and may have contributed to the diversification of feeding behaviour.

The advantage of a large foraging range is clear. According to the allometric scaling of 'foraging radius' (*sensu* Pennycuick 1979) with body size, natural selection should favour an increased foraging radius brought about by larger body size or a persistent fast gait (*e.g.*, animal proportions (mechanics) that maximise travel velocity). The latter would suggest a selection pressure for increased leg length, although stride properties at speed are strongly correlated with body size (Alexander *et al.* 1977), and so such adaptations would need to be independent of body size increases. Foraging velocity on the other hand is independent of body size as it is likely a result of perceptual, energetic and behavioural constraints (Shipley *et al.* 1996). Ideally a faster gait would be achieved at no greater net energy cost. However, during mammalian evolution, energy costs of locomotion have increased with body size (Taylor 1980), and relative to body size (*i.e.*, higher for larger animals relative to their mass (Underwood 1983)). And, while locomotion costs are quite small compared to total maintenance (*e.g.*, 8% of wildebeest (*Connochaetes taurinus*) energy intake is spent on travelling 10 km per day (Pennycuick 1979), although energy costs can increase four-fold on soft or waterlogged ground (Dijkman and Lawrence 1997)) this does imply a commensurate gain in energy intake for larger animals as a result of moving faster. The implication is that if it is too costly to travel to more abundant forage resources at distance from water sources, then there is an advantage in being able to efficiently utilize impacted resources more local to the water source.

Non-overlapping watering behaviour (Ayeni 1975) and assortative mating within groups specialising on vegetation at differing distances from water sources (Danley and Kocher 2001) may account for diversification stability under these conditions of shared geography. Studies of watering behaviours in a broad range of southern African species further suggest mechanisms for reinforcement including territoriality (Ritter and Bednekoff 1995), preference for drinking sites (Jarman 1972), avoidance of competition (Hitchcock 1996) and avoidance of larger or dominant species (Peters 1983, Parker 1997, Valeix *et al.* 2008).

ii. Allopatric speciation

Persistent patches, whether they comprise a single watering point or a network of watering points within travelling range, act as attractors to animal populations. A parallel may be drawn between island biogeography and the captive influence of dry-season, drinking-water location. Both may give rise to reproductive isolation as it is plausible that animals frequenting one of two patches may be isolated from animals in the other patch, if the patches are sufficiently separate (Fig. 2). For this to occur, the waterless distance between patches would need to be further than the possible travel between drinking events, and would therefore depend on species water dependency and mobility. Although models of allopatric speciation already include desert barriers, this case defines geographic isolation as a vicariance event in terms of a physiological barrier to dispersal and gene flow[3], independent of other environmental factors, for dry-season conditions over an evolutionary time scale.

iii. Parapatric speciation

Even if watering points are dispersed, they may not be sufficiently isolated to form discrete patches, but reinforcement mechanisms (see i. above) may act in parallel to further reinforce sympatric speciation by animal populations frequenting different watering points. A special case of this partial isolation would be where patches containing water sources are aligned along a boundary of a waterless zone, or along another geographic barrier (Fig. 3). Under these conditions populations frequenting adjacent watering points may mate, however intraspecies differences accumulated along the length of the species' geographic range are sufficiently large to preclude mating by the populations at its extremes, and differences along the length of the geographic range may be reinforced by the hybrid zones acting as genetic barriers (Barton and Hewitt 1985). These systems of populations distributed along habitat boundaries are collectively known as "ring species" (*e.g.*, *Ensatina eschscholtzii*, Wake *et al.* 1986).

---

[3] Darwin (1859) only refers to drinking at one point, early on in *The Origin of Species*, but he does identify the importance of water distribution towards a mechanism for speciation later, albeit as a physical barrier, and not a physiological barrier.

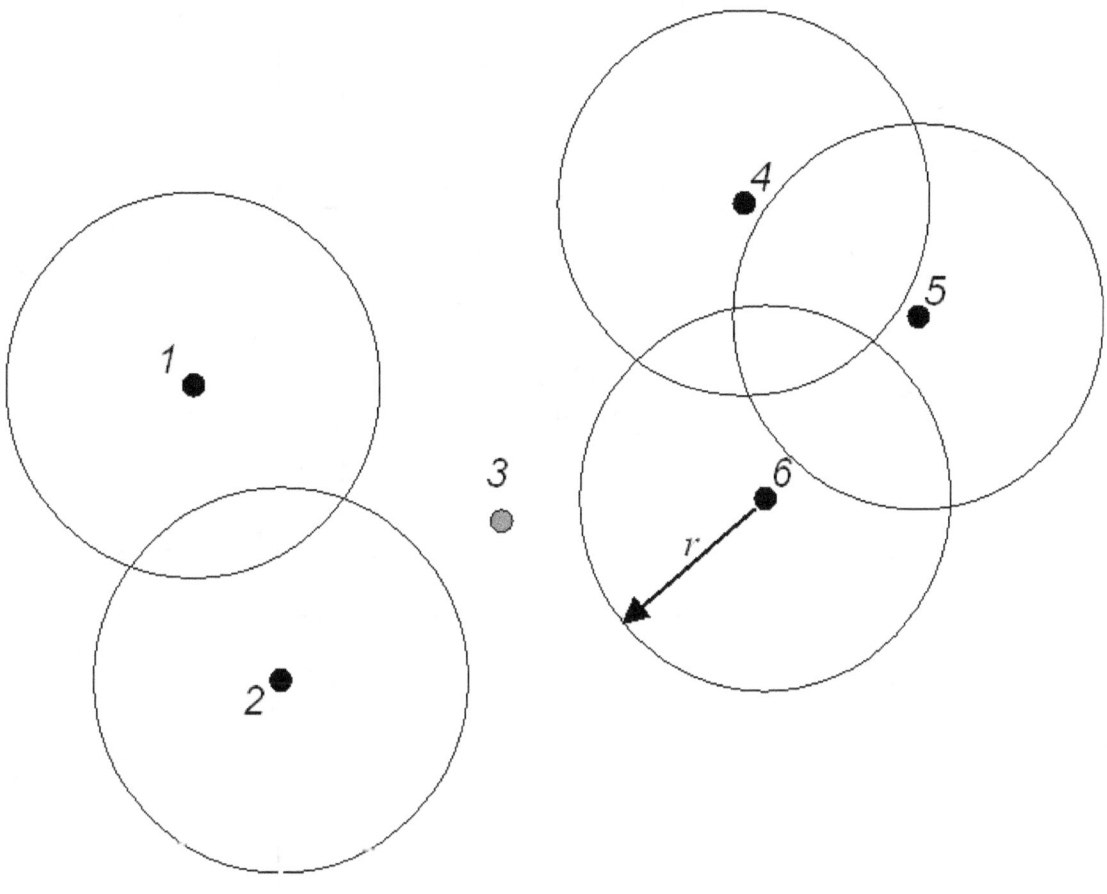

Fig. 2: A conceptual model of how populations using the group of two watering points at *1* and *2* can become isolated from animals using the group of watering points at *4*, *5* and *6*, as the watering point at *3* dries up. Foraging radius (r) is the furthest an animal species is able to travel before needing to return to water. r defines the extent of the piosphere associated with each watering point but is dependent on the numbers in each animal species (*e.g.*, individual animal sizes and group aggregation) and their impacts, plus environmental factors such as topography. Distances at which a broad range of modern, water-dependent animal species have been sighted from water during the dry season are typically reported below 10 km (*e.g.*, Ayeni 1975, Western 1975, Cumming and Cumming 2003).

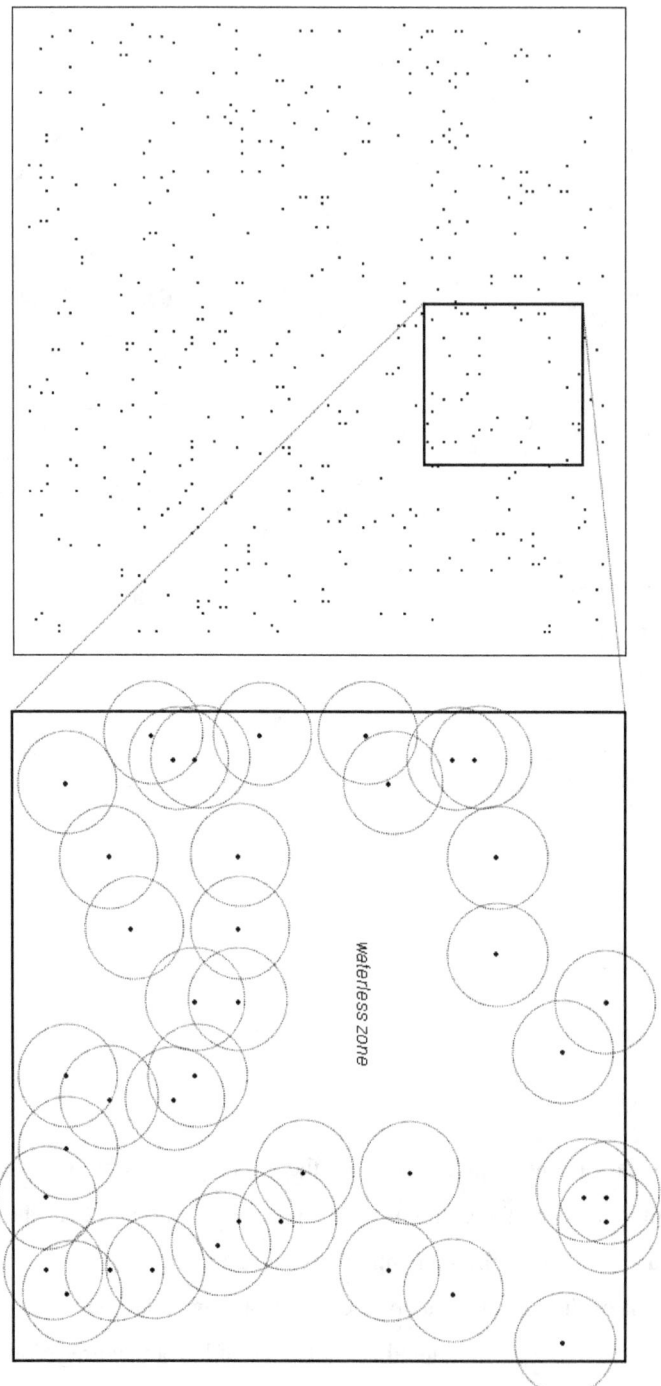

Fig. 3: A random distribution of points to illustrate this conceptual model of how a system of watering points organised along a habitat boundary (*waterless zone*) coupled with habitual utilisation of watering points by animal subpopulations could give rise to a ring species system (*inset*). Mating opportunities are indicated by overlaps in the spatial extent of piospheres from neighbouring watering points. This form of geographic separation leads to accumulated intraspecies differences between subpopulations and the formation of separate species.

Tests and predictions of the hypothesis

To test these different hypothetical models of speciation, it will be necessary to distinguish between habitat-related adaptations, utilisation gradient adaptations and water dependency isolations. While bovid postcranial fossils have been used widely to infer African paleohabitats (Kappelman *et al.* 1997, Reed 1998, DeGusta and Vrba 2003, Mendoza *et al.* 2005), genetic variation could be investigated between animal populations spanning a known geographical barrier, such as the Rift Valley. The hypothesis that water distribution can lead to allopatric speciation by isolating populations is amenable to testing by comparisons of gene flow among populations of water-dependent and water-independent species; gene flow between populations of the former would be expected to be lower (Fig. 4). Habitat fragmentation would not be expected to cause this difference. Microsatellite technology (*e.g.*, Marshal *et al.* 1998) would be ideal for this study, having recently been used to investigate spatial isolation in a population of red deer (*Cervus elaphus*, Nussey *et al.* 2005). Using multiple variable loci has also been shown to improve the assumptions underlying tests of species divergence, particularly those regarding complete reproductive isolation (Carstens and Knowles 2007).

Most useful perhaps is the opportunity to map genetic diversity and structure across the whole continent for large mammalian grazers, as already achieved for mixed-feeding bushbuck (*Tragelaphus scriptus*) (Moodley and Bruford 2007). Some understanding of the role of water dependency may result from comparing these distributions to records of habitat and geology. Even water dependency itself can be estimated for extinct species: dental enamel contains differing amounts of ^{18}O that make it possible to distinguish between the watering behaviours of grazers and browsers (Sponheimer and Lee-Thorp 1999, Kingston and Harrison 2007), with variable success for mixed feeders (Bibi 2007).

With respect to the case of persistent lineages in wet refugia (Vrba 1987), sizeable populations may have been distributed along the course of a network of water sources that enabled free mixing, as would have been possible for river valley corridors (Bobe 2006). Other smaller populations reliant on fewer water sources would have been more susceptible to extinction from drought, which should lead one to the obvious conclusion that fossil assemblages deposited after drought-related mass mortalities ought to be found in the vicinity of ancient water sources, as has already been found by Cohen *et al.* (2007). Furthermore, data on the habitual use of watering points by distinct groups of animals (philopatry) may already exist and parentage tests would go further towards investigating this potential of watering points in maintaining animal subpopulations.

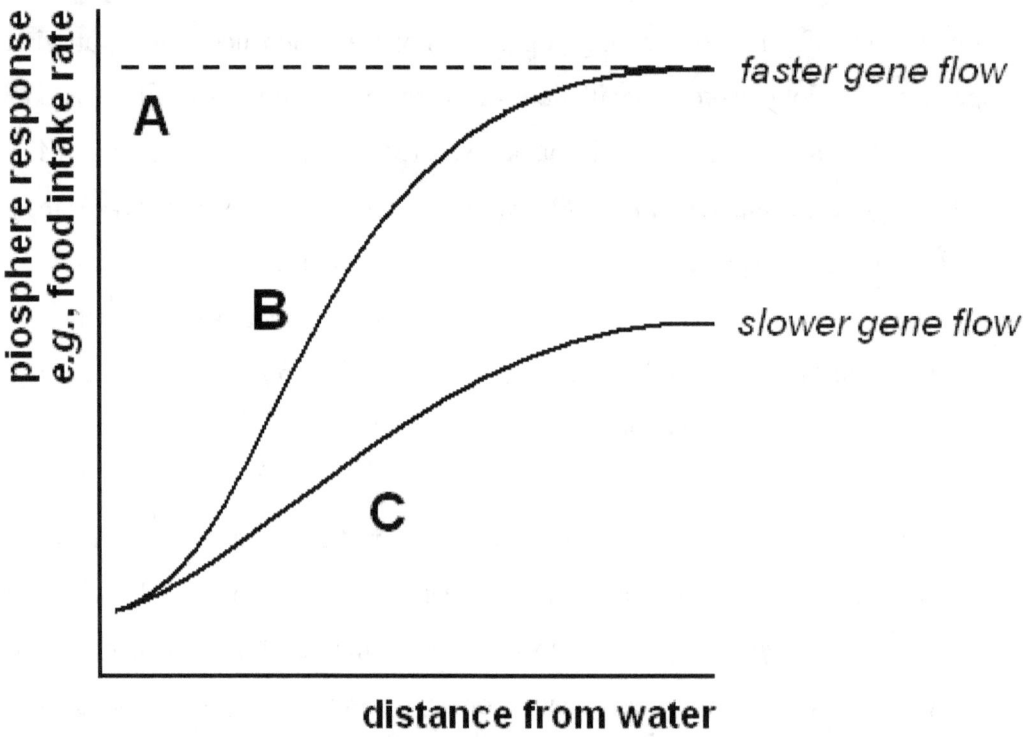

Fig. 4: Hypothetical piosphere response in forage supply (*e.g.*, food intake rate) for populations of animal species who are water-independent (A) and water-dependent (B and C), under normal climatic conditions (A and B) and drought (C). Water-independent animals (A) have little, or no, impact at water points. Water-dependent animals under normal conditions (B) impact forage resources which results in a utilization gradient. During droughts forage levels are further reduced along the length of the gradient (C). Gene flow would be expected to be lowest for water-dependent animals during droughts.

Adaptations under the sympatric model may be harder to distinguish from changes in response to general environmental conditions, but developments in quantitative palaeontology have revealed methods to identify dietary specializations from tooth morphology regardless of taxonomic affiliation (Jernvall *et al.* 1996, Mendoza and Palmqvist 2008). Usefully, feeding-type may also be unrelated to body size in dictating patterns of tooth wear (Veiberg *et al.* 2007) and this could provide an independent estimate of diet quality. It follows that cautious prediction of body mass using the craniodental morphology of extinct species could enable reliable estimates of body size independent of phylogeny and non-dietary, ecological adaptations (Mendoza *et al.* 2006).

Undoubtedly, body size is the single whole-animal indicator that best correlates to water requirement (Taylor 1968, Young 1970, Bothma 1996) and feeding strategy (Gordon and Illius 1996), whereas limb evolution, as long as it is independent of body size change (Scott 1985), is in response to the locomotory challenges encountered in different habitats (Kappelman 1988, Damuth and MacFadden 1990, Kappelman *et al.* 1997). So, mechanical adaptations should include: increased leg length for either faster travel or energetically cheaper travel (Christiansen 2002), and superior water conservation (*e.g.*, blood vessels for the counter-current system of blood cooling (Taylor 1972)). The fossil record may go some way to providing evidence for such adaptations whilst accounting for the alternative selective advantages of these traits (*e.g.*, increased leg length gives a better means of escape from predators): for example, it shows that cursorial grazers had long slender limbs (Jacobs 2004) and that all bovid hindlimb bones are more slender than those expected for their body sizes (Scott 1985).

The traditional interpretation of animal responses to rapid climatic fluctuations (*e.g.*, Potts and Deino 1995) suggests constraints on their distributions as imposed by biogeography (Lyons 2003), but could the predominant constraint on ranging be the distance a water-dependent animal can travel from their water sources? The hypothetical model proposed here defines the distances animals may travel away from water, before needing to return to drink, as the operational scale for vicariance. This modified model would position water location at the centre of our understanding of bovid evolution in Africa. As water resource specialists, in terms of the resource-use hypothesis, grazing ungulates would be predicted to incur more extinctions than browsing ungulates, as has been noted by Guthrie (1984, cited in Owen-Smith 1989). The traditional, unmodified model is unable to account for this because it does not incorporate the main difference critical for survival between animals of different feeding strategies, namely, water dependency. Frequenting a water source incurred the same daily risks of predation in the Pleistocene (Klein and Cruz-Uribe 1991) as it does today. But, undoubtedly the reason for catastrophic mass mortality of water-dependent populations in the past was very likely the same as it is today, the effect of drought on water sources and forage resources (but also see Owen-Smith 1989).

Indeed, there is growing evidence that animal populations are in equilibrium with their dry-season forage resources (Sæther 1997; Illius and O'Connor 1999, 2000; Derry 2004), but, instead of rainfall directly controlling vegetation production, the variability in food resources is more likely to be strongly coupled to changes in available foraging area, and this is moderated by the availability of drinking-water sources (Chamaille-Jammes *et al.* 2007). In simulations, piosphere extent has been shown to be positively correlated with animal density which operates with a lag matching the time taken to reach sexual maturity (Derry 2004).

The larger, mainly grazing species have adapted their birth seasons to coincide with the wet-season peak in their food supply (Sinclair *et al.* 2000). However, the availability of dry-season forage determines the rate of utilisation of body fat reserves generated during those wet seasons. Consequently, because forage resources are limited within the piosphere, compromised intake rates during the dry season reduce state-dependent reproductive success. As a result, piosphere extent is a function of animal density, and it exerts its influence at a different temporal scale, ultimately moderating population size independent of mortality rates (Derry 2004).

Finally, it is not impossible that the influence of climate has been overstated; re-evaluation of the North American fossil record has produced an alternative explanation that animal interaction is a more important driver of evolution than climate change (Alroy 2000). If the same were to be found for the African Pliocene-Pleistocene, this would suggest evolutionary mechanisms operating at a smaller scale than absolute habitat extent as defined by the dynamics of climate.

Conclusions

In summary, water dependency is likely to have played a major role in ungulate evolution. At the scale of the single piosphere or network of piospheres, selection pressures are expected to promote alternative feeding strategies, morphological adaptations and avoidance behaviour. These selection pressures are present in every dry season. Over longer time periods, distinct clusters of watering points introduce spatial structure into gene flow between populations and present the same evolutionary opportunities as island biogeography. Critics of this water-oriented model may argue that only a small amount of gene flow is required between populations to prevent the speciation process. A short period of heavy rainfall within an otherwise extended arid period with the accumulation of surface water would allow migration between isolates. However, this is offset to an extent in the water-oriented model, as time to death from dehydration (<1 week (Nicholson 1985, Kay 1997)) is much less than from starvation (*e.g.*, 8 weeks (Walker *et al.* 1987)). This suggests that drinking-water location exerts a dominant influence on animal survival and reproduction, and that these roles of drinking-water location need to be incorporated into existing models of large mammalian herbivore evolution in Africa.

Acknowledgements

JFD is grateful to numerous colleagues for their interest in this topic, particularly Stuart Blackman, Martyn Murray, William Bond, Peter Scogings and Loeske Kruuk, and for financial assistance for travel in southern Africa received from The Bath & West Agricultural Society and the James Rennie Bequest from the School of Biological Sciences at the University of Edinburgh. We also thank two anonymous referees for their comments on an earlier draft of the manuscript.

References

Abule E, Smit GN and Snyman HA 2005. The influence of woody plants and livestock grazing on grass species composition, yield and soil nutrients in the Middle Awash Valley of Ethiopia. *Journal of Arid Environments* 60: 343–358.

Adams J 2000. *A quick background to the Pliocene* [online]. Environmental Sciences Division, Oak Ridge National Laboratory. Available from http://www.esd.ornl.gov/projects/qen/pliocene.html [Accessed 14/03/08].

Adolph EF 1943. *Physiological Regulations*. Cattell Press, Lancaster, PA.

Alexander RMcN, Langman VA and Jayes AS 1977. Fast locomotion of some African ungulates. *Journal of Zoology, London* 183: 291-300.

Alroy J 2000. New methods for quantifying macroevolutionary patterns and processes. *Paleobiology* 26:707-733.

Andrew MH 1988. Grazing impact in relation to livestock watering points. *Trends in Ecology and Evolution* 3: 336-339.

Andrew MH and Lange RT 1986. Development of a new piosphere in arid chenopod shrubland grazed by sheep. 1. Changes to the soil surface. *Australian Journal of Ecology* 11: 395-409.

Anthony NM, Johnson-Bawe M, Jeffery K, Clifford SL, Abernethy KA, Tutin CE, Lahm SA, White LTJ, Utley J, Wickings EJ and Bruford MW 2007. The role of Pleistocene refugia and rivers in shaping gorilla genetic diversity in central Africa. *Proceedings of the National Academy of Sciences of the United States of America* 104: 20432-20436.

Aristotle 350BC. *Historia Animalium* (On the History of Animals). Available from http://classics.mit.edu/Aristotle/history_anim.html [Accessed 14/03/08].

Ayeni JSO 1975. Utilization of waterholes in Tsavo National Park (East). *East African Wildlife Journal* 13: 305-323.

Ayeni JSO 1977. Waterholes in Tsavo National Park, Kenya. *Journal of Applied Ecology* 14: 369-378.

Barnosky AD 2005. Effects of Quaternary climatic change on speciation in mammals. *Journal of Mammalian Evolution* 12: 247-264.

Barnosky AD, Koch PL, Feranec RS, Wing SL and Shabel AB 2004. Assessing the causes of late Pleistocene extinctions on the continents. *Science* 306: 70–75.

Barton NH and Hewitt GM 1985. Analysis of hybrid zones. *Annual Review of Ecology and Systematics* 16: 113–148.

Bauer IE, McMorrow J, Yalden DW 1994. The historic ranges of three Equid species in northeast Africa - a quantitative comparison of environmental tolerances. *Journal of Biogeography* 21: 169-182.

Beadle LC 1981. *The Inland Waters of Tropical Africa: An Introduction to Tropical Limnology*. Longman, London.

Beerling DJ and Osborne CP 2006. The origin of the savannah biome. *Global Change Biology* 12: 2023-2031.

Behrensmeyer AK, Todd NE, Potts R and McBrinn GE 2003. Late Pliocene faunal turnover in the Turkana Basin, Kenya and Ethiopia. *Science* 278: 1589-1594.

Bibi F 2007. Dietary niche partitioning among fossil bovids in late Miocene C_3 habitats: Consilience of functional morphology and stable isotope analysis. *Palaeogeography, Palaeoclimatology, Palaeoecology* 253: 529-538.

Bisigato AJ, Bertiller MB, Ares JO and Pazos GE 2005. Effect of grazing on plant patterns in arid ecosystems of Patagonian Monte. *Ecography* 28: 561-572.

Bobe R 2006. The evolution of arid ecosystems in eastern Africa. *Journal of Arid Environments* 66: 564-584.

Bobe R and Eck GG 2001. Responses of African Bovids to Pliocene Climatic Change. *Paleobiology* 27: 1-47.

Bonnefille R 1984. Palynological research at Olduvai Gorge. *National Geographic Society Research Papers* 17: 227-243.

Bothma J du P 1996. *Game Ranch Management*. J.L. van Schaik Uitgewers, Pretoria.

Braack L 1997. *A revision of parts of the management plan for the Kruger National Park, Volume VII: An objectives hierarchy for the management of the KNP*. National Parks Board, South Africa.

Brits J, van Rooyan MW and van Rooyan N 2002. Ecological impact of large herbivores on the woody vegetation at selected watering points on the eastern basaltic soils in the Kruger National Park. *African Journal of Ecology* 40: 53-60.

Carstens BC and Knowles LL 2007. Shifting distributions and speciation: species divergence during rapid climate change. *Molecular Ecology* 16: 619-627.

Chamaillé-Jammes S, Fritz H and Murindagomo F 2007. Climate-driven fluctuations in surface-water availability and the buffering role of artificial pumping in an African savanna: Potential implication for herbivore dynamics. *Austral Ecology* 32: 740–748.

Christiansen P 2002. Locomotion in terrestrial mammals: the influence of body mass, limb length and bone proportions on speed. *Zoological Journal of the Linnean Society* 136: 685–714.

Clauss M, Frey R, Kiefer B, Lechner-Doll M, Loehlein W, Polster C, Rössner GE, Streich, WJ 2003. The maximum attainable body size of herbivorous mammals: morphophysiological constraints on foregut, and adaptations of hindgut fermenters. *Oecologia* 136: 14-27.

Codron D, Lee-Thorp JA, Sponheimer M and Codron J 2007a. Nutritional content of savanna plant foods: implications for browser/grazer models of ungulate diversification. *European Journal of Wildlife Research* 53: 100-111.

Codron D, Lee-Thorp JA, Sponheimer M, Codron J, de Ruiter D and Brink JS 2007b. Significance of diet type and diet quality for ecological diversity of African ungulates. *Journal of Animal Ecology* 76: 526-537.

Cohen AS, Stone JR, Beuning KRM, Park LE, Reinthal PN, Dettman D, Scholz CA, Johnson TC, King JW, Talbot MR, Brown ET and Ivory SJ 2007. Ecological consequences of early Late

Pleistocene megadroughts in tropical Africa. *Proceedings of the National Academy of Sciences of the United States of America* 104: 16422-16427.

Coyne JA and Orr HA 2004. *Speciation*. Sinauer Associates, Sunderland, MA.

Cumming DHM and Cumming GS 2003. Ungulate community structure and ecological processes: body size, hoof area and trampling in African savannas. *Oecologia* 134: 560–568.

Damuth J and MacFadden BJ 1990. *Body size in mammalian paleobiology: estimation and biological implications*. Cambridge: Cambridge University Press.

Danley PD and Kocher TD 2001. Speciation in rapidly diverging systems: lessons from Lake Malawi. *Molecular Ecology* 10: 1075–1086.

Darwin C 1859. *The Origin of Species by means of Natural Selection; or, the Preservation of Favoured Races in the Struggle for Life*. John Murray, London.

DeGusta D and Vrba ES 2003. A method for inferring paleohabitats from the functional morphology of bovid astragali. *Journal of Archaeological Science* 30: 1009-1022.

deMenocal PB 2004. African climate change and faunal evolution during the Pliocene-Pleistocene. *Earth and Planetary Science Letters* 220: 3-24.

deMenocal PB and Bloemendal J 1995. Plio-Pleistocene climatic variability in subtropical Africa and the paleoenvironment of hominid evolution: a combined data-model approach. In: *Paleoclimate and Evolution with Emphasis on Human Origins*. Vrba ES, Denton GH, Partridge TC and Burckle LH (eds.). Yale University Press, New Haven. pp 262–288.

Derry JF 2004. *Piospheres in semi-arid rangeland: consequences of spatially constrained plant-herbivore interactions*. Unpublished PhD thesis. University of Edinburgh. Available from http://hdl.handle.net/1842/600 [Accessed 14/03/08].

de Wit M and Stankiewicz J 2006. Changes in surface water supply across Africa with predicted climate change. *Science* 311: 1917-1921.

Dijkman JT and Lawrence PR 1997. The energy expenditure of cattle and buffaloes walking and working in different soil conditions. *Journal of Agricultural Science, Cambridge* 128: 95-103.

du Toit JT and Cummings DHM 1999. Functional significance of ungulate diversity in African savannas and the ecological implications of the spread of pastoralism. *Biodiversity and Conservation* 8: 1643–1661.

du Toit PCV, Blom CD and Immelman WF 1995. Diet selection by sheep and goats in the arid Karoo. *African Journal of Range and Forage Science* 12: 16-26.

Edroma EL 1989. The response of tropical vegetation to grazing and browsing in Queen Elizabeth National Park, Uganda. *Symposia of the Zoological Society of London* 61: 1-13.

Egeland C 2007. *Zooarchaeological and Taphonomic Perspectives on Hominid and Carnivore Interactions at Olduvai Gorge, Tanzania*. Unpublished PhD thesis. Indiana University. Available from http://www.paleoanthro.org/dissertations/Charles%20Egeland.pdf [Accessed 14/03/08].

Essop MF, Harley EH and Baumgarten I 1997. A molecular phylogeny of some bovidae based on restriction-site mapping of mitochondrial DNA. *Journal of Mammalogy* 78: 377-386.

Fernández MH and Vrba ES 2005. Macroevolutionary processes and biomic specialization: testing the resource-use hypothesis. *Evolutionary Ecology* 19: 199–219.

Fernandez-Gimenez ME and Allen-Diaz B 1999. Testing a non-equilibrium model of rangeland vegetation dynamics in Mongolia. *Journal of Applied Ecology* 36: 871-885.

Fretwell SD and Lucas HL Jr 1970. On territorial behaviour and other factors influencing habitat distribution in birds. I. Theoretical development. *Acta Biotheoretica* 19: 16-36.

Friedel MH 1988. The development of veld assessment in the Northern Transvaal Savanna. II. Mixed bushveld. *Journal of the Grassland Society of Southern Africa* 5: 55-63.

Fritz H, de Garine-Wichatitsky M and Letessier G 1996. Habitat use by sympatric wild and domestic herbivores in an African savanna woodland: the influence of cattle spatial behaviour. *Journal of Applied Ecology* 33: 589-598.

Gagnon M and Chew AE 2000. Dietary Preferences in Extant African Bovidae. *Journal of Mammalogy* 81: 490-511.

Gasse F 2006 Climate and hydrological changes in tropical Africa during the past million years. *Comptes Rendus Palevol* 5: 35-43.

Gaylard A, Owen-Smith RN and Redfern J 2003. Surface water availability: Implications for heterogeneity and ecosystem processes. In: *The Kruger Experience: Ecology and Management of Savanna Heterogeneity.* du Toit JT, Rogers KH and Biggs HC (eds.). Island Press, Washington. pp 171-188.

Georgiadis NJ, Kat PW, Oketch H and Patton J 1990. Allozyme divergence within the Bovidae. *Evolution* 44: 2135-2149.

Getzin S 2005. The suitability of the degradation gradient method in arid Namibia. *African Journal of Ecology* 43: 340–351

Gillooly JF, Allen AP, West GB and Brown JH 2005. The rate of DNA evolution: effects of body size and temperature on the molecular clock. *Proceedings of the National Academy of Sciences of the United States of America* 102: 140-145.

Gordon IJ and Illius AW 1996. The nutritional ecology of African ruminants: a reinterpretation. *Journal of Animal Ecology* 65: 18-28.

Goudie AS 1996. Climate: Past and Present. In: *The Physical Geography of Africa.* Adams WM, Goudie AS and Orme AR (eds.). Oxford University Press, Oxford. pp 34-59.

Graetz RD and Ludwig JA 1978. A method for the analysis of piosphere data applicable to range assessment. *Australian Rangeland Journal* 1: 126-136.

Grove AT, Street FA and Goudie AS 1975. Former lake levels and climatic change in the Rift Valley of Southern Ethiopia. *Geographical Journal* 141: 177-194.

Guthrie RD 1984. Mosaics, allelochemics and nutrients: an ecological theory of late Pleistocene megafaunal extinctions. In: *Quaternary extinctions*. Martin PS and Klein RG (eds.). University of Arizona Press, Tucson. pp 259-298.

Guthrie RD 2006. New carbon dates link climatic change with human colonization and Pleistocene extinctions. *Nature* 441: 207-209.

Henly SR, Ward D and Schmidt I 2007. Habitat selection by two desert adapted ungulates. *Journal of Arid Environments* 70: 39-48.

Heshmatti GA, Facelli JM and Conran JG 2002. The piosphere revisited: plant species patterns close to waterpoints in small, fenced paddocks in chenopod shrublands of South Australia. *Journal of Arid Environments* 51: 547–560.

Hitchcock D 1996. Wildlife observed in Kutse Game Reserve, Botswana, at pans with either artificial or natural water sources. *African Journal of Ecology* 34: 70-74.

Hoffman RR 1989. Evolutionary steps of ecophysiological adaptation and diversification of ruminants: a comparative view of their digestive system. *Oecologia* 78: 443-457.

Hudson RJ 1985. Body size, energetics and adaptive radiation. In: *Bioenergetics of Wild Herbivores*. Hudson RJ and White RG (eds.). CRC Press, Inc. Florida. pp 1-24.

Illius AW and Gordon IJ 1991. Prediction of intake and digestion in ruminants by a model of rumen kinetics integrating animal size and plant characteristics. *Journal of Agricultural Science* 116: 145-157.

Illius AW and Gordon IJ 1992. Modelling the nutritional ecology of ungulate herbivores: evolution of body size and competitive interactions. *Oecologia* 89: 428-434.

Illius AW and O'Connor TG 1999. On the relevance of nonequilibrium concepts to arid and semiarid grazing systems. *Ecological Applications* 9: 798-813.

Illius AW and O'Connor TG 2000. Resource heterogeneity and ungulate population dynamics. *Oikos* 89: 283-294.

Jacobs BF 2004. Paleobotanical studies from tropical Africa: relevance to the evolution of forest, woodland, and savannah biomes. *Philosophical Transactions of the Royal Society, B* 359: 1573-1583

Janis CM 1989. A climatic explanation for patterns of evolutionary diversity in ungulate mammals. *Palaeontology* 32: 463-481.

Janis CM 2008. An evolutionary history of browsing and grazing ungulates. In: *The Ecology of Browsing and Grazing*. Gordon IJ and Prins HHT (eds.). Ecological Studies 195. Springer Berlin Heidelberg. pp 21-45.

Janis CM, Damuth J and Theodor JM 2002a. The origins and evolution of the North American grassland biome: the story from the hoofed mammals. *Palaeogeography, Palaeoclimatology, Palaeoecology* 177: 183-198.

Janis CM, Theodor JM and Boisvert B 2002b. Locomotor evolution in camels revisited: a quantitative analysis of pedal anatomy and the evolution of the pacing gait. *Journal of Vertebrate Paleontology* 22: 110-121.

Jarman PJ 1972. The use of drinking sites, wallows and salt licks by herbivores in the flooded Middle Zambezi Valley. *East African Wildlife Journal* 10: 193-209.

Jones CG, Lawton JH and Shachak M 1994. Organisms as ecosystem engineers. *Oikos* 69: 373-386.

Jernvall J, Hunter JP and Fortelius M 1996. Molar tooth diversity, disparity, and ecology in Cenozoic ungulate radiations. *Science* 274: 1489-1492.

Johnson TC, Scholz CA, Talbot MR, Kelts K, Ricketts RD, Ngobi G, Beuning K, Ssemanda I and McGill JW 1996. Late Pleistocene dessication of Lake Victoria and rapid evolution of cichlid fishes. *Science* 273: 1091-1093.

Kappelman J 1988. Morphology and locomotor adaptations of the bovid femur in relation to habitat. *Journal of Morphology* 198: 119-130.

Kappelman J, Plummer T, Bishop L, Duncan A and Appleton S 1997. Bovids as indicators of Plio-Pleistocene paleoenvironments in East Africa. *Journal of Human Evolution* 32: 229-256.

Kay RNB 1997. Responses of African livestock and wild herbivores to drought. *Journal of Arid Environments* 37: 683-694.

Kimura M and Ohta T 1969. The average number of generations until fixation of a mutant gene in a finite population. *Genetics* 61: 763-771.

Kingdon J 2003. *Lowly Origins: When, Where, and Why Our Ancestors First Stood Up*. Princeton University Press, Princeton, NJ.

Kingston JD and Harrison T 2007. Isotopic dietary reconstructions of Pliocene herbivores at Laetoli: Implications for hominin paleoecology. *Palaeogeography, Palaeoclimatology, Palaeoecology* 243: 272-306.

Klein RG and Cruz-Uribe K 1991. The bovids from Elandsfontein, South Africa, and their implications for the age, paleoenvironment, and origins of the site. *African Archaeological Review* 9: 21–79.

Knight MH 1995. Drought-related mortality of wildlife in the southern Kalahari and the role of man. *African Journal of Ecology* 33: 377-394.

Knowles LL and Richards CL 2005. Importance of genetic drift during Pleistocene divergence as revealed by analyses of genomic variation. *Molecular Ecology* 14: 4023–4032.

Köhler M and Moyà-Solà S 2001. Phalangeal adaptations in the fossil insular goat *Myotragus*. *Journal of Vertebrate Paleontology* 21:621-624.

Kumar S and Subramanian S 2002. Mutation rates in mammalian genomes. *Proceedings of the National Academy of Sciences of the United States of America* 99: 803-808.

Lailhacar S, Mansilla A, Faundez L and Tonini P 1993. The piosphere effect of a goat corral on the productivity of arid mediterranean-type rangelands in northern Chile. *Proceedings of the XVII International Grassland Congress* 17: 73-75.

Lancaster N 1989. Late Quaternary Palaeoenvironments in the South-western Kalahari. *Palaeogeography, Palaeoclimatology, Palaeoecology* 70: 367-376.

Lange RT 1969. The piosphere, sheep track and dung patterns. *Journal of Range Management* 22: 396-400.

Lyons SK 2003. A quantitative assessment of the range shifts of Pleistocene mammals. *Journal of Mammalogy* 84: 385-402.

MacPhee RDE and Marx PA 1998. *Lightning Strikes Twice: Blitzkrieg, Hyperdisease, and Global Explanations of the Late Quaternary Catastrophic Extinctions*. American Museum of Natural History.

Martin PS and Klein RG (eds.) 1984. *Quaternary extinctions: a prehistoric revolution*. University of Arizona Press, Tucson.

McNaughton SJ and Georgiadis NJ 1986. Ecology of African Grazing and Browsing Mammals. *Annual Review of Ecology and Systematics* 17: 39-66.

Mendoza M, Janis CM and Palmqvist P 2005. Ecological patterns in the trophic-size structure of large mammal communities: a 'taxon-free' characterization. *Evolutionary Ecology Research* 7: 505–530

Mendoza M, Janis CM and Palmqvist P 2006. Estimating the body mass of extinct ungulates: a study on the use of multiple regression. *Journal of Zoology* 270: 90–101.

Mendoza M and Palmqvist P 2008. Hypsodonty in ungulates: an adaptation for grass consumption or for foraging in open habitat? *Journal of Zoology* (OnlineEarly Articles).

Moleele NM 1994. *Ecological change and piospheres: Can the classical range succession model and its modification explain changes in vegetation and soil around boreholes in eastern Botswana?* Unpublished MSc. Thesis, University of Canberra.

Moodley Y and Bruford MW 2007. Molecular biogeography: towards an integrated framework for conserving pan-African biodiversity. *PLoS ONE* 2:e454.

Mphinyane WN 2001. *Influence of livestock grazing within piospheres under free range and controlled conditions in Botswana*. Unpublished PhD Thesis, University of Pretoria.

Murray MG and Illius AW 1996. Multispecies grazing in the Serengeti. In: *The ecology and management of grazing systems*. Hodgson J and Illius AW (eds.). CAB International, Oxon, UK. pp 247-272.

Nangula S and Oba G 2004. Effects of artificial water points on the Oshana ecosystem in Namibia. *Environmental Conservation* 31: 47–54.

Nash MS, Whitford WG, de Soyza AG, Van Zee JW and Havstad KM 1999. Livestock Activity and Chihuahuan Desert Annual-Plant Communities: Boundary Analysis of Disturbance Gradients. *Ecological Applications* 9: 814-823.

Nicholson MJ 1985. The water requirements of livestock in Africa. *Outlook on Agriculture* 14: 156-164.

Nussey DH, Coltman DW, Coulson T, Kruuk LEB, Donald A, Morris SJ, Clutton-Brock TH and Pemberton J 2005. Rapidly declining fine-scale spatial genetic structure in female red deer. *Molecular Ecology* 14: 3395–3405.

Odling-Smee FJ, Laland KN and Feldman MW 2003. *Niche Construction: The Neglected Process in Evolution*. Princeton University Press.

Owen-Smith RN 1987. Pleistocene Extinctions: The Pivotal Role of Megaherbivores. *Paleobiology* 13: 351-362.

Owen-Smith RN 1988. *Megaherbivores: The influence of very large body size on ecology*. Cambridge University Press, Cambridge.

Owen-Smith RN 1989. Megafaunal Extinctions: The Conservation Message from 11,000 Years B.P. *Conservation Biology* 3: 405-412.

Owen-Smith RN 1996. Ecological guidelines for waterpoints in extensive protected areas. *South African Journal of Wildlife Research* 26: 107-112.

Parker G 1997. *Animal interactions at waterholes in a region of high elephant density*. Unpublished BSc. dissertation, University of Edinburgh.

Pennycuick CJ 1979. Energy costs of locomotion and the concept of "foraging radius". In: *Serengeti: Dynamics of an ecosystem*. Sinclair ARE and Norton-Griffiths M (eds.). University of Chicago Press, Chicago. pp 164-184.

Pérez-Barbería FJ, Elston DA, Gordon IJ and Illius AW 2004. The evolution of phylogenetic differences in the efficiency of digestion in ruminants. *Proceedings of the Royal Society of London B* 271: 1081-1090.

Pérez-Barbería FJ, Gordon IJ and Pagel M 2002. The origins of sexual dimorphism in body size in ungulates. *Evolution* 56: 1276-1285.

Perkins JS 1991. *The impact of borehole dependent cattle grazing on the environment and society of the eastern Kalahari sandveld, Central District, Botswana*. Unpublished PhD thesis. University of Sheffield, UK.

Perkins JS 1996. Botswana: fencing out the equity issue. Cattleposts and cattle ranching in the Kalahari Desert. *Journal of Arid Environments* 33: 503-517.

Perkins JS and Thomas DSG 1993. Spreading deserts or spatially confined environmental impacts? Land degradation and cattle ranching in the Kalahari desert of Botswana. *Land Degradation and Rehabilitation* 4: 179-194.

Peters RH 1983. *The Ecological Implication of Body* Size. Cambridge University Press, Cambridge, UK.

Pienaar D, Biggs H, Deacon A, Gertenbach W, Joubert S, Nel F, van Rooyan L and Venter F 1997. A revised water-distribution policy for biodiversity maintenance in the KNP. In: *A Revision of Parts of the Management Plan for the Kruger National Park. Volume VIII: Policy Proposals Regarding Issues Relating to Biodiversity Maintenance, Maintenance of Wilderness Qualities and Provision of Human Benefits*. Braack L (ed.). South African National Parks. Unpublished Report, Skukuza, South Africa. pp. 165–200.

Potts R and Deino A 1995. Mid-Pleistocene change in large mammal faunas of East Africa. *Quaternary Research* 43: 106–113.

Prins HHT 1996. *Ecology and behaviour of the African buffalo*. Chapman & Hall, London.

Prins HHT 1998. Origins and development of grasslands in northwestern Europe. In: *Grazing and Conservation Management*. WallisDeVries MF, Bakker JP and Van Wieren SE (eds.). pp 55-105.

Prins HHT and Fritz H 2008. Species Diversity of Browsing and Grazing Ungulates: Consequences for the Structure and Abundance of Secondary Production. In: *The Ecology of Browsing and Grazing*. Gordon IJ and Prins HHT (eds.). Ecological Studies 195. Springer Berlin Heidelberg. pp 179-200.

Raia P and Meiri S 2006. The island rule in large mammals: Paleontology meets ecology. *Evolution* 60: 1731-1742.

Redfern JV, Grant R, Biggs H and Getz WM 2003. Surface-water constraints on herbivore foraging in the Kruger National Park, South Africa. *Ecology* 84, 2092–2107.

Reed KE 1998. Using Large Mammal Communities to Examine Ecological and Taxonomic Structure and Predict Vegetation in Extant and Extinct Assemblages. *Paleobiology* 24: 384-408.

Reid C, Slotow R, Howison O and Balfour D 2007. Habitat changes reduce the carrying capacity of Hluhluwe-Umfolozi Park, South Africa, for Critically Endangered black rhinoceros *Diceros bicornis*. *Oryx* 41: 247-254.

Ritter RC and Bednekoff PA 1995. Dry season water, female movements and male territoriality in springbok: Preliminary evidence of water-hole-directed sexual selection. *African Journal of Ecology* 33: 395-404.

Rogers KH and O'Keefe J 2003. River heterogeneity: Ecosystem structure, function, and management. In: *The Kruger Experience: Ecology and Management of Savanna Heterogeneity*. du Toit JT, Rogers KH and Biggs HC (eds.). Island Press, Washington. pp 189-218.

Roux C 2006. *Feeding ecology, space use and habitat selection of elephants in two enclosed game reserves in the Eastern Cape Province, South Africa*. Masters thesis, Rhodes University.

Ryan SJ and Jordaan W 2005. Activity patterns of African buffalo *Syncerus caffer* in the Lower Sabie Region, Kruger National Park, South Africa. *Koedoe* 48: 117-124.

Ryan SJ and Getz WM 2005. A spatial location–allocation GIS framework for managing water sources in a savanna nature reserve. *South African Journal of Wildlife Research* 35: 163–178.

Sæther B-E 1997. Environmental stochasticity and population dynamics of large herbivores: a search for mechanisms. *Trends in Ecology and Evolution* 12: 143-149.

Sankaran M, Hanan NP, Scholes RJ, Ratnam J, Augustine DJ, Cade BS, Gignoux J, Higgins SI, Le Roux X, Ludwig F, Ardo J, Banyikwa F, Bronn A, Bucini G, Caylor KK, Coughenour MB, Diouf A, Ekaya W, Feral CJ, February EC, Frost PGH, Hiernaux P, Hrabar H, Metzger KL, Prins HHT,

Ringrose S, Sea W, Tews J, Worden J and Zambatis N 2005. Determinants of woody cover in African savannas. *Nature* 438: 846-849.

Schoener TW 1974. Resource partitioning in ecological communities. *Science* 185: 27-39.

Scott KM 1985. Allometric trends and locomotor adaptations in the Bovidae. *Bulletin of the American Museum of Natural History* 179: 197–288.

Sepulchre P, Ramstein G, Fluteau F, Schuster M, Tiercelin J-J, Brunet M 2006. Tectonic uplift and eastern African aridification. *Science* 313: 1419-1423.

Sinclair ARE 1983. The adaptations of African ungulates and their effects on community function. In: *Tropical Savannas*. Bourlíere F (ed.). Tropical Savannas, Ecosystems of the World. Volume 13. Elsevier, Amsterdam. pp 401-426.

Sinclair ARE and Fryxell JM 1985. The Sahel of Africa: ecology of a disaster. *Canadian Journal of Zoology* 63: 987-994.

Sinclair ARE, Mduma SAR and Arcese P 2000. What determines phenology and synchrony of ungulate breeding in Serengeti? *Ecology* 81: 2100–2111.

Shipley LA, Spalinger DE, Gross JE, Hobbs NT and Wunder BA 1996. The dynamics and scaling of foraging velocity and encounter rate in mammalian herbivores. *Functional Ecology* 10: 234-244.

Shrader AM, Owen-Smith N, Ogutu JO 2006. How a mega-grazer copes with the dry season: food and nutrient intake rates by white rhinoceroses in the wild. *Functional Ecology* 20:376-384

Smet M and Ward D 2006. Soil quality gradients around water-points under different management systems in a semi-arid savanna, South Africa. *Journal of Arid Environments* 64: 251–269.

Smit IPJ, Grant CC, Devereux BJ 2007. Do artificial waterholes influence the way herbivores use the landscape? Herbivore distribution patterns around rivers and artificial surface water sources in a large African savanna park. *Biological Conservation* 136: 85-99.

Sponheimer M and Lee-Thorp JA 1999. The ecological significance of oxygen isotopes in enamel carbonate. *Journal of Archaeological Science* 26: 723–728.

Stebbins GL 1981. Coevolution of grasses and herbivores. *Annals of the Missouri Botanical Garden* 68: 75-86.

Stokes CJ and Yeaton RI 1994. A line-based vegetation sampling technique and its application in succulent karoo. *African Journal of Range Forage Science* 11: 11-17.

Street-Perrott FA and Gasse F 1981. Recent development in research into the Quaternary Climatic History of the Sahara. In: *The Sahara: Ecological Change and Early Economic History*. Allen JA (ed.). Longman, London. pp 7-28.

Taylor CR 1968. The minimum water requirements of some East African bovids. *Symposium of the Zoological Society of London* 21: 195-206.

Taylor CR 1972. The desert gazelle: a paradox resolved. *Symposium of the Zoological Society of London* 31: 215-227.

Taylor CR 1980. Energetics of locomotion: primitive and advanced mammals. In: *Comparative Physiology: Primitive Mammals*. Schmidt-Nielsen K, Bolis L and Taylor CR (eds.). Cambridge University Press. pp 192-199.

Thomas DSG and Shaw PA 1991. *The Kalahari Environment*. Cambridge University Press, Cambridge.

Thomas DSG and Shaw PA 2002. Late Quaternary environmental change in central southern Africa: new data, synthesis, issues and prospects. *Quaternary Science Reviews* 21: 783-797.

Thrash I 1997. Infiltration rate of soil around drinking troughs in the Kruger National Park, South Africa. *Journal of Arid Environments* 35: 617-625.

Thrash I and Derry JF 1999. The nature and modelling of piospheres: a review. *Koedoe* 42: 73-94. Pretoria.

Thrash I, Nel PJ, Theron GK and Bothma J du P 1991. The impact of the provision of water for game on the basal cover of herbaceous vegetation around a dam in the Kruger National Park. *Koedoe* 34: 121-130.

Thrash I, Theron GK and Bothma J Du P 1993. Impact of water provision on herbaceous plant community composition in the Kruger National Park. *African Journal of Range and Forage Science* 10: 31-35.

Tolkamp BJ, Allcroft DJ and Kyriazakis I 1999. Estimating meal criteria for meal pattern analysis of dairy cows. In: *Proceedings of the British Society of Animal Science* Winter Meeting 1999. British Society of Animal Science, Penicuik, UK. p 203.

Traill LW 2004. Seasonal utilization of habitat by large grazing herbivores in semi-arid Zimbabwe. *South African Journal of Wildlife Research* 34: 13 – 24.

Tyler JA and Hargrove WW 1997. Predicting spatial distribution of foragers over large resource landscapes: a modeling analysis of the Ideal Free Distribution. *Oikos* 79: 376-386.

Underwood R 1983. Feeding behaviour of grazing African ungulates. *Behaviour* 84: 195-243.

UNEP 2002. *Vital Climate Graphics Africa: The Impacts of Climate Change*. United Nations Environment Programme, UNEP/GRID-Arendal.

Valeix M, Fritz H, Matsika R, Matsvimbo F and Madzikanda H 2008. The role of water abundance, thermoregulation, perceived predation risk and interference competition in water access by African herbivores. *African Journal of Ecology* (OnlineEarly Articles).

Van der Schijff HP 1957. *'n Ekologiese studie van die flora van die Nasionale Krugerwiltuin*. D.Sc. Thesis, Potchefstroom Universiteit vir Christelike Hoër Onderwys, Potchefstroom.

van Heezik Y, Ismail K and Seddon PJ 2003. Shifting spatial distributions of Arabian oryx in relation to sporadic water provision and artificial shade. *Oryx* 37: 295-304.

van Hooft WF, Groen AF and Prins HHT 2003. Genetic structure of African buffalo herds based on variation at the mitochondrial D-loop and autosomal microsatellite loci: Evidence for male-biased gene flow. *Conservation Genetics* 4: 467-477.

Van Sickle J, Attwell CAM, Craig GC 1987. Estimating Population Growth Rate from an Age Distribution of Natural Deaths. *Journal of Wildlife Management* 51: 941-948.

Veiberg V, Loe LE, Mysterud A, Solberg EJ, Langvatn R and Stenseth NC 2007. The ecology and evolution of tooth wear in red deer and moose. *Oikos* 116: 1805–1818.

Vermeij GJ 1989 Arguments on Evolution: A Paleontologist's Perspective by Antoni Hoffman. *Paleobiology* 15: 199-203.

Vrba ES 1980. Evolution, species and fossils: how does life evolve? *South African Journal of Science* 76: 61–84.

Vrba ES 1987. Ecology in relation to speciation rates: some case histories of Miocene-Recent mammal clades. *Evolutionary Ecology* 1: 283–300.

Vrba ES 1992. Mammals as a key to evolutionary theory. *Journal of Mammalogy* 73: 1–28.

Vrba ES and DeGusta D 2004. Do species populations really start small? New perspectives from the Late Neogene fossil record of African mammals. *Philosophical Transactions of the Royal Society, B* 359: 285-293.

Wake DB, Yanev KP and Brown CW 1986. Intraspecific sympatry in allozymes in a "ring species," the plethodontid salamander *Ensatina eschscholtzii*, in southern California. *Evolution* 40: 866-868.

Walker BH, Emslie RH, Owen-Smith RN and Scholes RJ 1987. To cull or not to cull: lessons from a southern African drought. *Journal of Applied Ecology* 24: 381-401.

Washington-Allen RA, Van Niel TG, Ramsey RD and West NE 2004. Remote sensing–based piosphere analysis. *GIScience and Remote Sensing* 41: 136-154.

Weir JS 1971. The effect of creating additional water supplies in a Central African National Park. In: *The scientific management of animal and plant communities for conservation.* Duffey E and Watt AS (eds.). Blackwell Scientific Publications, Oxford. pp. 367-385.

Western D 1975. Water availability and its influence on the structure and dynamics of a savannah large mammal community. *East African Wildlife Journal* 13: 265-286.

Wilmshurst JF, Fryxell JM and Colucci PE 1999. What constraints daily intake in Thomson's gazelles? *Ecology* 80: 2338–2347.

WoldeGabriel G, Haile-Selassie Y, Renne PR, Hart WK, Ambrosek SH, Asfaw B, Heiken G and White T 2001. Geology and palaeontology of the Late Miocene Middle Awash valley, Afar rift, Ethiopia. *Nature* 412: 175-178.

Young E 1970. *Water as faktor in die ekologie van wild in die Nasionale Krugerwildtuin.* Unpublished DSc Thesis. University of Pretoria, SA.

Zambatis N 1985. The relationships between evaporation and certain physical parameters of circular pans and rectangular troughs. *Koedoe* 28: 87-92.

Zimov SA, Chuprynin VI, Oreshko AP, Chapin FS, Reynolds JF and Chapin MC 1995. Steppe-tundra transition: a herbivore-driven biome shift at the end of the pleistocene. *The American Naturalist* 146: 765-94.

Penis size and Peacocks' Eyes

150 years ago, Darwin first discusses sexual selection in *The Origin*... then comprehensively in *Descent of Man*. The problem was why some characters seem non-adaptive and not governed by natural selection. A famous example is the tail of the male peacock, who Darwin noticed desperately seeks "a spectator of some kind", and, "will shew off his finery before poultry, or even pigs".

What advantage this behaviour has for survival is not obvious: the cost to the male in producing his stunning apparel mustn't exceed any survival benefits but the way that Darwin carefully constructs a considered and logical argument demonstrates his exemplary working. He shows that the benefit is real and great; not so much involved with individual survival, it is related to the reason for surviving. Reproduction.

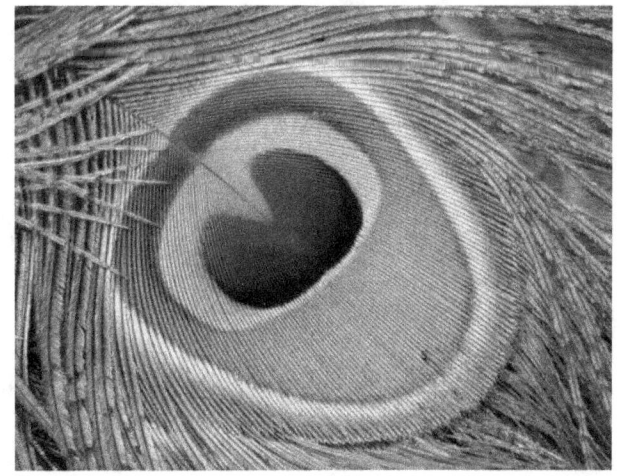

Darwin cleverly identified the benefit that outweighs the cost of cumbersome and garish plumage, also predicting some optimal balance in the economy of nature. Costs may define the fine line between getting it right and "going all Liberace", but the converse of underselling is worse: peahens are a lot fussier than Darwin thought. Recently, the number of 'eye-spots' were adjusted without shortening the tail; typically there are about 150 of them, shown to be optimal. Less, and mating success dwindled until males with fewer than 130 had less chance of pulling a peahen than Quasimodo with a feather duster stuck up his arse. In fact, girl power dictates that even two dusters might not be enough; there is an inbuilt, self-perpetuating inflation that drives up costs, but also profit margins. This positive feedback loop is called "Fisherian Runaway" after R.A. Fisher who described it (1915, 1930). For the peahen, choosing a mate who has a beautiful long tail increases the chance of her male offspring growing one, and so also be attractive. Of course, for her there is very good reason to be choosy. Males can and want to spray their sperm around willy-nilly, but producing an egg and offspring levies a prohibitive tax upon her metabolism.

Whilst researching the evolutionary ecology of African antelopes I couldn't fail to notice the sexually selective adornments that they boast. Very horny! From the captivating spiral of the Kudu to the seductive Arabian curve of the Giant Sable's scimitar, it's advertising with a full page spread. Of course, take out too many column inches like the Irish Elk, and nature's economy becomes bankrupt, a recession destined for extinction. But get it right, and … well, there was an impala I knew with forty females and territory the size of a nature reserve. He was called "Lucky Boy".

Darwin could only sketch out the details of how sexual selection works. As with natural selection and genetics, he couldn't have been privy to the peacock's immune system, the health of which is strongly linked to his tail. This was also discovered recently, but is likely to be duplicated in other species. For example, stick an extension onto a long-tailed widow bird and other than ending up with a longer-tailed widow bird, you'll also gain a queue of fawning females. Impressive plumage makes a great advert for virility, because if there's no parasites depleting your immune system, then you should be a healthy specimen. Of course, it is this underlying adaptive advantage that is really being passed on to any offspring. But there's more for the female than just genetic imperative. Spotting a brightly-coloured, prancing dandy is adaptive; it saves time searching for a mate reducing her costs and increasing survival.

Avoiding jokes about size mattering, for males given a short straw, there is hope. When costs in choosing are low, females don't necessarily prefer only one male ornament, selecting between those on offer irrespective of the costs to the males in their production. Again, our  peacock's tail length is an indicator of antibody production while the eye-spots are a measure of phagocytic defences. Sometimes this doesn't work as well as it might. Embellishments exhibit varied responses to an affliction. The female may then be drawn by an unaffected attribute, costing her dearly if she's actually selected a pox-ridden impostor. Fisher's process can produce multiple preferences and the males must follow by ensuring their bits are preferred, whatever and to whomever the price. Rutting and lekking males, like Lucky Boy and Darwin's peacocks and stags, are good examples of this, strutting their various stuffs. But it can also be introduced artificially: preferring beaks and plumage, female zebra finches also like flashy leg bands, like wearing a Rolex.

Similarly, Darwin had a lot to say about human adornment, specifically beards. He traced the roots of facial hair to "an ancient progenitor" which was then acted on by sexual selection. Yet again modern studies prove that he was right. There is a strong sexual connotation to facial hair because males use it to advertise their social rank and virility. Like muscle mass and penis development, facial hair growth is influenced by testosterone levels, and is like wearing a sign on your chin saying, "hair today, gonads tomorrow", which must make the World Beard & Moustache champion the sexiest man in the world. At the time of writing, his name is Mr Passion. Oddly enough, occidental society has recently tended away from beards, associating them with antisocial psyches or disturbed pasts, as in "hiding behind a beard".

Darwin's own beard has achieved iconic status, the bearded version being more readily called to the minds of most than the young man. However, my own personal experiences provide little support for sexual selection of male facial hair. I have at times grown a beard in various styles, its presence more dependent on my pennies than my penis, and I'm sure it hardly ever contributed carnally. Clearly it was never long enough. I am now habitually clean-shaven, but I wonder what private thoughts Darwin was entertaining as he sat there in his dotage looking like old Father Time. I mean, his was so long that he could have draped it over a shoulder. Not necessarily his own.

Wars of the Word

A blog at least gives people the 'right of reply'. It is rare for a newspaper to 'waste' copy real estate for the purpose, and comments at the bottom of a page are often a jungle of misinformed misdirection, paraphrasing and obfuscation.

It's not surprising then, when not given a proper forum for discourse to be voiced openly, review and revenge must piggyback on mechanisms already out there, and bitterness can become shrouded in double-talk and camouflaged ambush. Of course, when the online forum is entirely public and moderated by a third or remote party, then those websites are more susceptible to this kind of abuse. *Amazon* is a good example.

Earlier this year, in what is evidently, "a row that has scandalised the academic world", Orlando Figes (Professor of History at Birkbeck College, University of London, *above left*), falsely posted highly critical reviews on *Amazon* of books written by some of his fellow historians: of Rachel Polonsky's (*above middle*) *Molotov's Magic Lantern* he wrote that it was, "the sort of book that makes you wonder why it was ever written", and of Robert Service's (*above right*) *Comrades* that it "is an awful book. It is very poorly written and dull …". Figes' own works he found strangely, "Beautifully written ... leaves the reader awed, humbled, yet uplifted ... with his superb story-telling skills. I hope he writes for ever". Figes is clearly enamoured of himself. Unfortunately, the court wasn't and he has since been ordered to pay damages and costs after a messy denial and exposure, and a subsequent libel hearing, and now finds himself with all the writing time he needs, on extended "sick leave".

Another high profile *Amazon* review-enacted spat occurred even more recently. Philip Kerr (*above left*) seemingly objected to reviews appearing in sequential years in *The Scotsman*, for his *A Quiet Flame* ("enjoyable enough, good-quality airport fiction. But that's all it is") and *If the Dead Rise Not* ("This is fiction to pass the time"). As both reviews were written by Allan Massie (*above right*), Kerr used the *Amazon* review section associated with Massie's *The Royal Stuarts: A History of the Family That Shaped Britain* to lambast it and much of his previous publication history ("reads like a chapbook for one of Massie's worthy but mostly unread novels"), adding "... Massie has reviewed my last two novels with a distinct lack of enthusiasm ... I'm of the opinion that authors should avoid reviewing books of their peers and, usually, I stick to this principle, but I've made a special exception ...". Massie (who, I can confirm from my time at the Wigtown Book Festival recently, is as witty in real life)

retorted, "For a professional writer to do this without a fee is agreeably unusual".

It's nothing new of course, professional rivalry between creatives probably started when cave-painters Ug accused Og of copying his mammoth's tail, and has been raging ever since. A little later, Charles Darwin (*above left*) also fell foul of a fellow author, Richard Owen (*above right*), who reviewed his *Origin of Species*. Here is an excerpt from my Darwin in Scotland that recounts their falling out:

> One of the most damning accusations that can be made of a scientist is plagiarism. By the time the Origin of Species was published, Owen already had been caught falsely claiming the discovery of a type of 'belemnite', an extinct group of molluscs closely related to modern squids. This serious offence cost Owen his places on the councils of both the Zoological Society and the Royal Society. Now, Owen was well known to be anti-evolution. So, when it came to his anonymous review in the Edinburgh Review of the Origin of Species in 1860, in which he suggested, in the third person, to have usurped Darwinian evolution by a full 10 years, the Darwinians were understandably outraged, particularly Huxley, Owen's despised adversary over human origins. In contrast to his dealings with the abandoned Grant, when Owen attacked Darwinian evolution as presented in the Origin of Species, he found himself confronted by what essentially constituted a team of Darwin's supporters. His ruse hadn't worked, even though, for the

purposes of subterfuge, Owen had fiendishly incorporated three of his own recent works, which he then proceeded to review, but was actually attempting to present evidence of an ongoing development in his evolutionary ideas:

> In his last published work Professor Owen does not hesitate to state 'that perhaps the most important and significant result of palæontological research has been the establishment of the axiom of the continuous operation of the ordained becoming of living things' [...] As to his own opinions regarding the nature or mode of that 'continuous creative operation', the Professor is silent. He gives a brief summary of the hypotheses of others, and as briefly touches upon the defects in their inductive bases. Elsewhere he has restricted himself to testing the idea of progressive transmutation by such subjects of Natural History as he might have specially in hand: as, e.g. the characters of the chimpanzee, gorilla, and some other animals.

Darwin, confused by the actions of his old friend, expresses deep hurt and humility when describing events to Asa Gray (1810–1888), his most ardent supporter in America:

> Have you seen how I have been thrashed by Owen in last Edinburgh: he misquotes & misrepresents me badly, & how he lauds himself. But the manner in which he sneers at Hooker is scandalous, to speak of his Essay & never allude to his work on Geograph. Distribution is scandalous. When Hooker's Essay appeared Owen wrote a note, which I have seen, full of strongest praise! What a strange man he is. All say his malignity is merely envy because my Book has made a little noise. How strange it is that he can be envious about a naturalist, like myself, immeasurably his inferior! But it has annoyed me a good deal to be treated thus by a friend of 25 years duration. He might have been just as severe without being so spiteful. Owen consoles himself by saying that the whole subject will be forgotten in ten years.

In confidence to Lyell, Darwin was more matter-of-fact, but still quite incredulous of Owen's departures from reality:

I have very long interview with Owen, which perhaps you would like to hear about, but please repeat nothing. Under garb of great civility, he was inclined to be most bitter & sneering against me. Yet I infer from several expressions, that at bottom he goes immense way with us. He was quite savage & crimson at my having put his name with defenders of immutability. When I said that was my impression & that of others, for several had remarked to me, that he would be dead against me: he then spoke of his own position in science & that of all the naturalists in London, 'with your Huxleys', with a degree of arrogance I never saw approached. He said to effect that my explanation was best ever published of manner of formation of species. I said I was very glad to hear it. He took me up short, 'you must not at all suppose that I agree with it in all respects'. I said I thought it no more likely that I shd be right on nearly all points, than that I shd toss up a penny & get heads twenty times running.

I asked him which he thought the weakest parts, he said he had no particular objection to any part. He added in most sneering tone if I must criticise I shd say 'we do not want to know what Darwin believes & is convinced of, but what he can prove'. I agreed most fully & truly that I have probably greatly sinned in this line, & defended my general line of argument of inventing a theory, & seeing how many classes of facts the theory would explain. I added that I would endeavour to modify the 'believes' & 'convinceds'. He took me up short, 'You will then spoil your book, the charm of(!) it is that it is Darwin himself'. He added another objection that the book was too 'teres atque rotundus', that it explained everything & that it was improbable in highest degree that I shd succeed in this. I quite agree with this rather queer objection, & it comes to this that my book must be very bad or very good. Lastly I thanked him for Bear & Whale criticism, & said I had struck it out. 'Oh have you, well I was more struck with this than any other passage; you little know of the remarkable & essential relationship between bears & whales'.

I am to send him the reference, & by Jove I believe he thinks a sort of Bear was the grandpapa of Whales! I do not know whether I have wearied you with these details which do not repeat to any one. We parted with high terms of consideration; which on reflexion I am almost sorry for. He is the most astounding creature I ever encountered.

All Darwin could do was declare in resignation and with regret: 'The Londoners say he is mad with envy because my book is so talked about. It is painful to be hated in the intense degree with which Owen hates me'.

The latest bitter rivalries to make the headlines are also scientists, and the events that surrounded the squabble were equally historic: James Watson (*below left*) and Francis Crick's (*below middle left*) discovery of the DNA double helix: their original papers detailing the structure, *A Structure for Deoxyribose Nucleic Acid* and *Genetical Implications of the structure of Deoxyribonucleic Acid* are both from 1953. The online archive that holds these profoundly historic papers also has the complementary papers by Maurice Wilkins (*below middle right*), Alex Stokes and Herbert Wilson, and Rosalind Franklin (*below right*) and Raymond Gosling, all incredibly from the same year. The relations between those involved has been understood to have been fraught; can you imagine the reviews that might have been?

The tensions are now confirmed: some of Crick's personal correspondence, known to exist but missing for years and thought to have been destroyed, has recently turned up amongst the papers of an office-sharing colleague, Sydney Brenner, himself also a Nobel laureate for his work on "genetic regulation of organ development and programmed cell death".

The correspondence (spread across "nine archive boxes of correspondence, photographs, postcards, preprints, reprints, meeting programmes, notes and newspaper cuttings, dates from 1950 to 1976, the bulk from the mid-1950s to the mid-1960s") documents in more detail perhaps the pivotal, and certainly one of the first direct incidents that soured relations between the MRC Unit at King's College, London (Wilkins & Franklin), and The Cavendish Laboratory in Cambridge (Watson & Crick). And, once again, it involves the academic anathema of plagiarism.

The paper in Nature that describes the found letters, The lost correspondence of Francis Crick details how,

On 21 November 1951, Franklin described her latest results in a colloquium at King's. Watson attended but left mistaken over the amount of water in the DNA structure ... Watson's order-of-magnitude underestimate of the water content led Crick to believe that there were very few possible structures for DNA and the right one might be found through model building alone ... As soon as Franklin saw the model — a triple helix ... she knew it was wrong.

This debacle precipitated a moratorium on further DNA work for Watson and Crick, who were doing no experimental work of their own. By most accounts, John Randall, the head of the MRC unit at King's, and William Lawrence Bragg, his equivalent at the Cavendish, called this halt after a quiet chat. But the recovered papers reveal correspondence between Wilkins and Crick in parallel to — perhaps even in place of — direct communication between Randall and Bragg. Thus, on 11 December 1951 we find a typed letter from Wilkins to Crick ... :

My dear Francis, I am afraid the average vote of opinion here, most reluctantly and with many regrets, is against your proposal to continue the work on n.a. [nucleic acids] in Cambridge. An argument here is put forward to show that your ideas are derived directly from statements made in the colloquium and this seems to me as convincing as your own argument that your approach is quite out of the blue ... I think it most important that an understanding be reached such that all members of our laboratory can feel in future, as in the past, free to discuss their work and interchange ideas with you ... I have much to gain by discussing my own work with you and after your attitude on Saturday begin to have very slight uneasy feelings in this respect.

It is likely that Wilkins wrote this on behalf of Randall, hence the formal tones. However, he followed it up later the same day with a strikingly more comradely handwritten note,

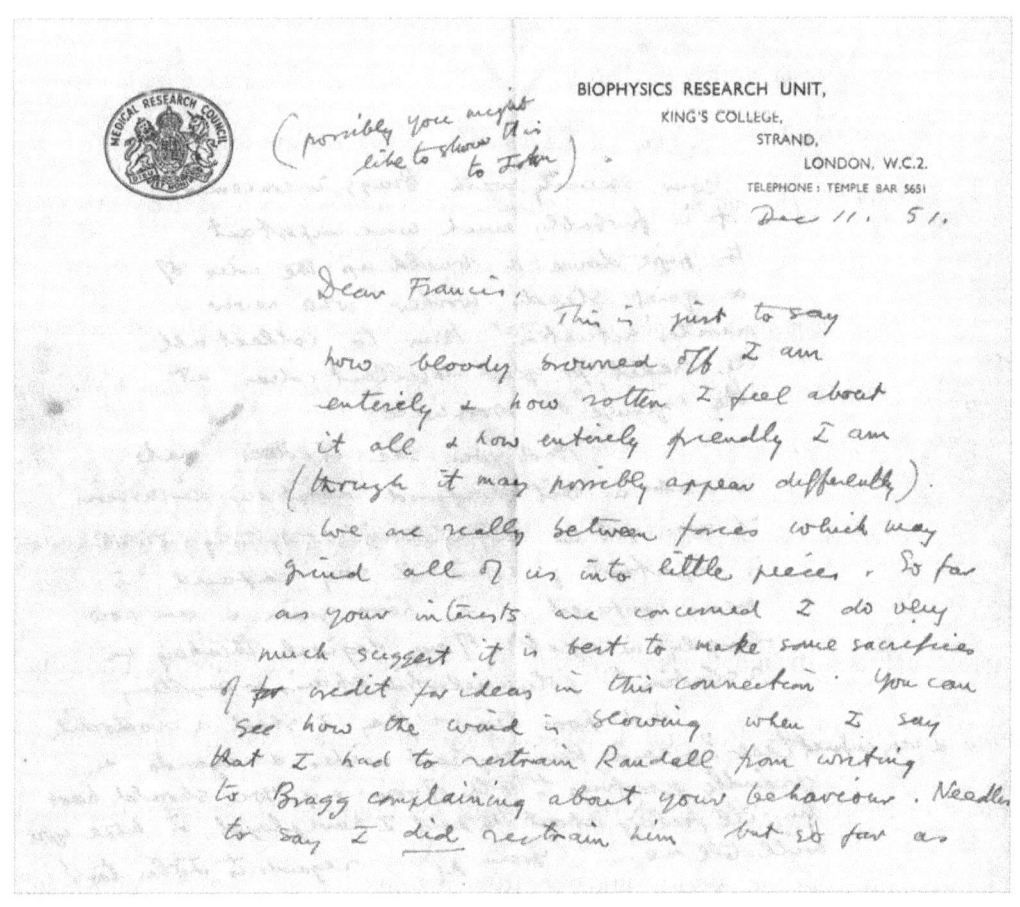

Dear Francis, This is just to say how bloody browned off I am entirely & how rotten I feel about it all & how entirely friendly I am (though it may possibly appear differently). We are really between forces which may grind all of us into little pieces ... I had to restrain Randall from writing to Bragg complaining about your behaviour. Needless to say I did restrain him, but so far as your security with Bragg is concerned it is probably much more important to pipe down & build up the idea of a quiet steady worker who never creates 'situations' than to collect all the credit for your excellent ideas at the expense of good will.

The team dynamics at King's were already fractured; Randall had omitted to formalise Franklin's role in the department, leaving Wilkins and Gosling to feel that their control over the diffraction studies of DNA had been usurped. Crick and Watson sympathised with their male counterparts, writing in response to Wilkins,

Dear Maurice, Just a brief note to thank you for the letters and to try to cheer you up. We think the best thing to get things straight is for us to send you a letter setting out in a mild manner our point of view. This will take a day or so to do, so we hope you'll excuse the delay. Please don't worry about it, because we've all agreed that we must come to an amicable arrangement ... so cheer up and take it from us that even if we kicked you in the pants it was between friends. We hope our burglary will at least produce a united front in your group! Yours ever Francis Jim

The abrasive Franklin and the shy Wilkins were never able to improve upon their relationship, even after Randall delegated their duties across the apparent two structures for DNA: the A- and B-forms. Wilkins favoured the B-form as it was likely helical. Franklin took the A-form and recorded results that suggested a non-helical structure, in contradiction to Wilkins who believed all DNA was helical. In what Wilkins no doubt thought of as a joke in very poor taste, Franklin and Gosling then presented him with a death notice for helical A-DNA:

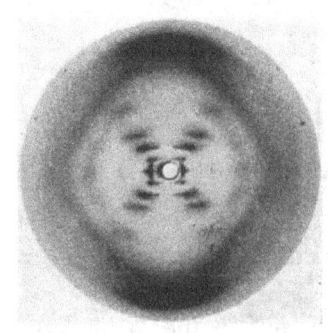

While Franklin was preparing to leave King's for Birbeck, Wilkins was still searching for definitive evidence of a helical structure for his B-form. So when, towards her leaving, Gosling passed him the famous photograph 51 that provided that evidence and had been taken by Franklin some time beforehand, Wilkins was understandably frustrated, confiding in Crick,

> *My dear Francis, ... The crystalline data is clearing up nicely. To think that Rosie had all the 3D data for 9 months & wouldn't fit a helix to it and there was I taking her word for it that the data was anti-helical. Christ. ...*

and later wrote again to Crick, looking to better times ahead,

> *... Let's have some talks afterwards when the air is a little clearer. I hope the smoke of witchcraft will soon be getting out of our eyes ...*

A week after this something happened that would dictate the future of biological science. On 30th January, 1953, Watson rushed to King's with a manuscript that had arrived a couple of days before, describing a structure for DNA by Linus Pauling.

There was a feeling of great competition with Pauling, and Watson and Crick knew his triple helix was wrong. So, when Watson couldn't locate Wilkins, he burst in upon Franklin to help disentangle Pauling's model.

No doubt still smarting from having previously been plagiarised by Watson, Franklin was uncooperative causing a scene. Watson backed away, straight into Wilkins who had come to investigate the commotion. Assessing the situation and likely biased by his own history with Franklin, Wilkins sided with Watson, revealing to him the all-important photograph 51. In this instance, animosity between colleagues had led to a rare positive outcome, and at what cost to Franklin?

If your confidence in the review system, and the peer-to-peer relations that may lead to its collapse, hasn't been dented enough already by these cases, then perhaps the funniest (on-topic) response to Service's self-serving [*ahem*] rant in the *Guardian, Comment is Free, The shame of Orlando Figes* may provide the final blow necessary, "… Grow a pair, for god's sake. We're talking about reviews on *Amazon* FFS. Nobody cares. There are eminent book publishers who keep my wife in a free supply of childrens' books in return for writing this guff. It's the way the world works". What a world, eh?

TO MARK THE BICENTENARY OF CHARLES DARWIN'S BIRTH AND THE 150TH ANNIVERSARY OF HIS SEMINAL WORK, 'ON THE ORIGIN OF SPECIES', WE ASKED FOUR EXPERTS

Does Genomics Need Darwin?

Well, does Genomics need Darwin? At first, it might seem like an overly simplistic question, deserving an obvious answer. But on closer inspection, a maze of convoluted ethics and morality is revealed lurking, darkly underneath.

Since the publication of On the Origin of Species, the science-religion divide arguably has polarised society. But there has always been another controversial axis skewering Darwin's ideas: one about which revolves Social Darwinism.

Hilter, Stalin, Mao, Pol Pot. Darwin? This historical connection between genocidal tyranny and the mild-mannered, Victorian gentleman naturalist is, of course, more tenuous than some anti-Darwinists assert. Nonetheless, Herbert Spencer was influenced by Darwinism. Even eugenics was given formal status and promoted from within Darwin's own family, notably by his cousin Sir Francis Galton.

Current controversies are no less com- plicated, nor less critical. Eugenicists advocate control of human reproduction through artificial selection of human characters. Today, technological developments have actuated this from a postpartum practice to an antenatal one. Foremost, modern genetics and genomics make the science fiction of designer babies a scientific reality.

And yet other recent developments, such as new insights into epigenetic mechanisms, suggest unpredictability within the genetic code, posing additional challenges for reproductive technologies and bioethics.

So, where does that leave our under- standing of Darwin's connection with genomics? Given the negative consequences of his contributions, one might even ask whether the Social Sciences ought to be celebrating him at all this year. Genomics, eugenics, Social Darwinism and designer babies: a set of terms that can be, and often are, used almost interchangeably to drag skeletons from the darkest depths of humanity's closeted past.

Here, we ask four social scientists to guide us through the relationship between Darwin, genomics and society. In considering one question each, they illuminate modern understandings of Darwin's ideas, and some of society's darker moral recesses.

Professor Brian Wynne
Associate Director, Cesagen

Why does Darwin's theory of evolution by natural selection remain controversial after 150 years?

Whether Darwin's theory of evolution by natural selection remains controversial today is itself a controversial matter. What that theory is, and what it is claimed to encompass, is not finally closed and precise.

Looking back to the 19th century context of Darwin's theory, it is worth noting what historians of science have demonstrated to be typical of scientific knowledge, namely that it is always invested seamlessly with different public meanings. This reflects scientists' and others' natural inclinations to find larger meaning in what may be more finite, albeit important, observations. Thus, what may be defined as 'scientific controversy', as if the controversy were solely and purely between competing scientific propositional beliefs with no wider social implication or interest, leaves an important ambiguity as to what kind of 'controversy' it is that we are considering. With Darwin and evolution, this ambiguity prevailed in the 19th century, as institutions of authority like the Church, aided and abetted by many anti-religious scientists, saw the theory of natural selection and its associated vision of human species biological continuity with the apes and 'lesser forms', as an assault on religious doctrines of the divine creation of Man. Many social philosophers like Herbert Spencer were also busy using Darwin's and nature's innocent name ideologically to justify right-wing policies of competitive 'nature red in tooth-and-claw'. Dogmatically entrenched belief engendered dogmatic response, even if entrenchment is gradual.

Thus, discriminating strictly 'scientific' conflict and maybe controversy from broader social equivalents is inevitably complex, and may be itself an unavoidably interpretive enterprise. The social- ist biologist and contemporary of Darwin, A R Wallace, reached similar conclusions to those of Darwin on evolution, but saw natural selection and 'survival of the fittest' as a function of primarily cooperative, not competitive, actions and relations. These differences were charged with much passion, thanks largely to two factors: first, the (different) wider meanings attached to what might have been strictly biological observations and theories, and second, that opposing or merely different stand- points, were seen as not just technically incorrect, but illegitimate in a moral sense, too. Such ambiguities, and moral investments in scientific knowledge, reign also in current times, and science itself is not immune from making them.

Nowadays, it would be folly to give scientific standing to the social controversies that persist around Creationism and its denial of the long-run history of human evolution from previous natural species. Yet if we ask about scientific understanding of the origins of life (as distinct from human life), we find important differences, which reflect differences as to what can properly be called 'scientific'.

Thus, Intelligent Design can be seen to have attempted to create controversy over contemporary scientific interpretations of Darwinian evolution, while care- fully separating itself from Creationism, which dogmatically literalises the book of Genesis story of Divine creation of humankind in six days. It is also careful to accept the scientific validity of Darwinian evolutionary theory, but challenges that evolution cannot explain the origins of life, since it cannot explain where life came from in the first place – evolutionary science can only explain the 'how?'s thereof, not the 'why?'s.

Whether this is a legitimate scientific challenge to science is an interesting question. If scientists want to claim that science already explains such matters, or will do sometime, then that hypothesis is a defensible (if arguable) one. But then so too is the challenge. Whether the hypothesis is scientific however, is itself a matter of dispute. It will never be simple to distinguish matters of 'scientific controversy' from matters of 'controversy' - and nowhere more so perhaps than around questions of life itself.

**Professor Hilary Rose Visiting Professor,
London School of Economics**

Given the history of Social Darwinism, are designer babies the new eugenics?

Charles Darwin was unquestionably a great scientist and crucial for the development of the life sciences, but we celebrate his birthday as if he were the only theorist of evolution, an essay unlikely to endear us to the rest of the West, to say nothing of historians of Islamic science conscious of scholars writing about natural selection 800 years earlier. But Darwin's theory of natural and sexual selection has long had multiple readings. Thomas Huxley publicly and controversially insisted that the diversity of species from one common origin made God redundant. By the end of the 20th century the Pope himself was confident that belief in God and evolution were entirely compatible. Natural scientists, particularly biologists, think that Herbert Spencer's Social Darwinism was a perverse reading of Darwinian theory, while those from the social studies of science, conscious of Darwin's debt to Malthus, are less convinced of the distinction.

But it was Galton's study of inherited genius and his science of eugenics, both endorsed by Darwin, which introduced the idea that artificial selection employed by animal and plant breeders should be applied to the human population. Welcomed by intellectuals, both progressive and conservative politicians and pro-birth-control feminists, only Catholics were outside the growing consensus. In England Darwin's son Leonard became chair of the Eugenics Society. Different eugenic practices developed: the British and Dutch preferred incarceration of the unfit in sexually segregated institutions while the Scandinavians saw compulsory sterilisation of the unfit as essential for the developing Welfare State. While many believe that eugenics stopped with the defeat of the Nazis, only the word fell out of favour, eugenics practices continued in most of Europe and the US until the mid seventies. With the rise of the women's movement, State eugenics died out.

By the 1980s, molecular genetics began to locate gene sequences associated with inherited disease. The Human Genome Project also served to fan genetic reductionism concerning behaviour, suggesting anything from alcoholism, sexual orientation and criminality to voting preference as being determined by our genes. But the HGP's promised gene therapy turned out to be more hype than hope. Diagnostics outstripped therapeutics. The HGP's claims also saw the dawn of consumer eugenics. Now it was the pregnant woman, in a culture that emphasises the overwhelming value of a healthy normal child, who had to decide which foetus to abort and which to keep.

And yet as the disability movement has radically challenged the belief that every woman wants a healthy normal baby; for a woman to discover that her foetus has a dodgy sequence does not necessarily mean that she will feel compelled to abort. Probably most will decide to terminate if the condition is severe, but the evidence that more children with Down's syndrome are being born suggests the word eugenics, meaning the well born, is hopefully becoming richer and more inclusive.

Given the special needs of children with disabilities, whether this more generous spirit will survive the recession remains to be seen.

Dr Staffan Müller-Wille
Research Fellow, Egenis

Would the Human Genome Project have happened without Darwin's theory of evolution by natural selection?

This is a tricky question. The relationship between Darwin's theory of evolution and the achievements of molecular biology over the past 50-odd years, culminating in the publication of a "working draft" sequence of the complete human genome in 2001, is far from straightforward. On the face of it, it is clearly possible to be a non-Darwinian and still hold that DNA is of special significance – a good example is Francis Collins, one of the leaders of the

Human Genome Project, who argues in his book The Language of God that "DNA sequence alone, even if accompanied by a vast trove of data on biological function, will never explain certain special human attributes, such as the knowledge of the Moral Law and the universal search for God."

On a more subtle level, detailed historical reconstructions have shown that molecular biology developed largely independently from evolutionary biology and even genetics. Molecular biology applies physical and chemical methods to elucidate the molecular mechanisms that underlie biological functions like inheritance, respiration, or cell-to-cell communication. Such an approach to phenomena of the living world does not have to take into account evolutionary history. As the philosopher of biology Paul Griffiths likes to put it, it is even advantageous for a molecular biologist to be a "methodological creationist" by holding that a good, mechanistic explanation of a function or process has been achieved once we can imagine the mechanism to have been designed by an engineer.

On the other hand it remains questionable if we could fully decipher the meaning of the multitude of patterns discernible in human DNA without having some grounding in Darwinian evolutionary theory. DNA carries with it signals from our evolutionary past which are wholly unaccountable for by theories of special creation. Thus Collins is more than happy to concede that "a common ancestor for humans and mice is virtually inescapable". When it comes to observing, interpreting and controlling such exceedingly complex processes as the spread of a contagious disease, it is even vital to take into account that the microbes causing the disease are subject to rapid evolutionary change, depending on spontaneous mutations and selective interaction with local environments, alternate hosts, and human populations. Diseases like bird flu have recently attracted much attention from epidemiologists endorsing this perspective.

If this is taken into account, there does appear to be a historical connection between Darwin's theory of evolution by natural selection and the discovery of DNA. Long before that discovery, the geneticist Herman J. Muller had become convinced that genes must be material particles capable of reproducing mutations faithfully once they had occurred. That DNA, with its peculiar structure of comple- mentary base sequences, was able to do exactly this, was one of the major reasons that convinced Watson and Crick in 1953 that they had solved "one of the fundamental biological problems."

Professor John Dupré
Director, Egenis

What light do new developments in molecular biology shed on the importance of Darwin's natural selection compared to, say, Lamarckian modes of inheritance?

The strong antipathy among evolutionists towards Lamarckianism (here under- stood very broadly as the idea that an organism can pass on characteristics acquired during its lifetime to its offspring) reached its peak in the mid- twentieth century, and was rooted in Weismann's doctrine of the separation of germline and soma, and the belief that only the latter was susceptible to the influence of the environment. More recently anti-Lamarckiansm has been grounded rather in the so-called Central Dogma of molecular biology, Francis Crick's hypothesis that information flowed only outwards from the DNA, to RNA, protein, and thence phenotypic features, but never back to the DNA. The kind of Lamarckianism that has most offended orthodox evolutionists would require violation of this principle through adaptive and heritable environ- mental effects on the DNA.

Extended views of inheritance, most obviously through models of cultural evolution, have long been recognized as introducing Lamarckian possibilities. But in fact there are paths of inheritance that bypass DNA even at the molecular level.

Reproduction involves the transmission from the mother of an entire cell, and this contains a great deal more than just DNA.

The significance of this is controversial. A common view has been that the cytoplasm contains just the mechanisms needed to translate the information in the DNA, the 'blueprint' or 'recipe' for the organism. However, recent discoveries have made this view increasingly untenable.

Genomes do different things in different cells in the body, and the same genetic sequences do different things in different organisms; much of the explanation of these differences lies in the cytoplasmic environment of the cell - both protein and RNA molecules interact with DNA sequences, promoting or inhibiting the transcription of particular sequences. Whereas until recently it was thought that the vast majority of the genome (>98%) not directly involved in coding for proteins was 'junk', perhaps selfish DNA involved in its own project of colonizing the genome, this view is now widely discredited. At least 70% of the genome appears to be transcribed, and it is increasingly suspected that much of this is involved in regulation of genome expression. Especially prominent among these regulatory elements are the small RNA molecules that are now known to be a major feature of all cellular environ- ments, and that have been divided into a rapidly diversifying taxonomy of kinds. Exploration of the functions of these systems of molecules is one of the most exciting areas of contemporary biology.

These developments don't yet show any need to rehabilitate Lamarckism. They do, however, indicate possible routes for Lamarckian processes, and recent work on epigenetics - modifications to the genome that affect its behaviour - give rapidly increasing plausibility to the belief that such processes really occur. A much-cited example is the research showing that maternal care in rats affects the behavioural dispositions of developing infant rats, and does so through a mechanism involving methy- lation - the best studied epigenetic process - of genes in the brain. Rats that have experienced this high quality maternal care are more disposed to provide high quality care to their own offspring, pointing to a mechanism for inheritance of the epigenetic effect that passes through behaviour. It remains a matter of speculation how widespread such processes are, but the possibility that there are major Lamarckian dimensions to evolution is now very much on the agenda.

Perhaps the only thing here that can be said with complete confidence is that ongoing developments in molecular biology will continue to transform significantly our view of evolutionary processes.

One Giant Leap for Mankind - Terrestrial and Extraterrestrial *Hominina* Evolution

[T]here is a striking parallelism in the laws of life throughout time and space.
                                     Charles Darwin, 1809-1882, *The Origin of Species.*[1]

In the business of futures, the modern-day soothsayer looks for definitive patterns and trends that may be extrapolated from the "now" to paint a better picture of what is going to happen next. While we humans (*Homo sapiens sapiens* of the subtribe *Hominina*) impact our planet to an increasing degree, the underlying laws of the physical world have always been more consistent in dictating those outcomes than any human-made influence. The human capacity for buffering ourselves from environmental effects only goes part of the way to divorcing ourselves from the natural world, Darwinian evolution, and the pressures of natural selection[2].

This paper first examines how human society has weakened our susceptibility to evolutionary pressures before looking forward to the most significant modern development, and the most likely cause of those pressures being remade stronger. Finally, predictions are made on the future of human evolution, on Earth and in space, including one scenario under which humanity is so externally controlled that selection occurs quite unnaturally.

Let us first consider how evolutionary processes may shape human societies under familiar conditions. This is quite a difficult question for us, mainly because Darwin's evolution really only deals with how individual forms arise, so how useful can it be when we want to look at the various interactive behaviours within larger groups of individuals? Look around you. You may be surrounded by others—in a library, coffee shop, office, school, or department. Did someone meet your gaze? You may be alone, in your own front room, but just beyond its portals I bet there are other homes nearby. You may even be in the car, experiencing some kind of Einsteinian hyper-relationship with the drivers of other cars around you; I wonder if that person who just swerved off the road and tipped into a ditch was also reading. Reading while driving is likely to get you nominated into the hallowed halls of the Darwin Awards.[3] It is doubtlessly dumb, and there is something funny about self-inflicted accidents. But, however condescending we feel about these unfortunates, many passersby would summon sufficient Samaritanism to stop and offer them assistance. It is the right thing to do. It is the humane thing to do.

For Darwin, such instincts were an inheritable trait, as susceptible to natural selection as any other, and he dedicated the whole of Chapter VII in *The Origin* to the exploration of the subject. The moral instinct may be ignored, however, just like the instincts to sleep, eat, and mate, which makes it debatable whether these behaviours can be considered instinctual at all. Nonetheless, philosopher and science historian Patrick Tort[4] explains how our morals have come to mold human society:

> [Darwin] has often been held responsible for the worst implementations of his theory to human societies such as "Social Darwinism," "neo-Malthusianism," eugenism, racism, brutal colonialism, ethnocide, or pro-slavery domination. However, Darwin was not only a staunch opponent to each of these movements, but he also gave the best theoretical arguments against them in the anthropological part of his works, especially in *The Descent of Man*.[5] Beyond being a peaceful philosopher, he certainly was the most convincing genealogist of *ethics*. [...]
>
> The *reversive effect* of evolution is the key concept in Darwinian anthropology. [...] The need for the reversive effect results from a paradox identified by Darwin in the

course of his attempt to extend to man the theory of *descent*, and from thinking about human morals and social future as a peculiar consequence and development of the former universal application of the selective law to the living.

The paradox can be put into words as follows: *Natural selection, as the ruling evolutionary principle, means elimination of the least fit in the struggle for life. Selection in humans leads to a social way of life, which rules that the more civilized you get, the more you tend to exclude the eliminatory behaviours through the interplay between morals and institutions. In a nutshell, natural selection selects civilization, which opposes natural selection.* How can we solve this problem?

We can do it easily by developing the very logic of the theory of selection itself. One of Darwin's fundamental points was that the natural selection selects not only organic variations showing the fittest adaptation, but also *instincts*. Among these advantageous (fittest) instincts, those Darwin calls the *social instincts* had been particularly chosen and developed, as confirmed by the universal triumph of a social way of life for mankind, and the tendency of "civilized" peoples towards hegemony. Now, in the state of civilization, which is the complex result of an increase of rationality, of the growing influence of the instinct of "sympathy," and of the several moral and institutional forms of altruism, can be observed a more and more systematic *reversal* of the individual and social behaviours as compared with what the mere continuation of the previous selective operation would be: Indeed, instead of the elimination of the less fit, civilization brings about the duty of assistance, which provides for them multiple actions of help and rehabilitation. Instead of the eventual demise of the sick and the disabled, it brings them protection, thanks to the operation of various new technologies and knowledge, like hygiene and medicine, [which] help them to survive and even minimize and compensate organic deficiencies. Acceptance of the destructive consequences of natural hierarchies of strength, of numbers, and of fitness for life have been replaced by a compensating interventionism which now attempts to suppress social disqualification.

Through *social instincts*, natural selection has selected *its opposite*. [...] Thus, the progressive emergence of *morals* seems to be a phenomenon inseparable from evolution, as a natural continuation of Darwinian materialism, and the inevitable extension from natural selection theory to the explanation of human society's future.

The popular understanding of this welfare provision is that human society has in part decoupled itself from the selective pressures of evolution. Within our society groups,

we can look after those that would otherwise not survive, unable to care for themselves, a basic requirement out there, in the rest of what we call "nature." We mostly don't die from cold, thanks to clothing, housing, and artificial sources of heat. We do not need to starve; our prey is readily available, arrayed before us on the supermarket shelves. Guns foil our predators and medicines cure our diseases. We have more than doubled our life expectancy with our technology and have largely modified our environment beyond recognition. We are the ultimate niche constructors.[6]

While human civilization has to a certain extent decoupled human beings from evolution, under the current range of conditions, warnings of changing climate and resource limitations suggest that conditions could become so variably extreme that our anthropogenic buffering from nature at best will be tested, at worst will fail. Even Nobel Prizes have been awarded to the global-warming doomsayers, the Intergovernmental Panel on Climate Change (IPCC) and Al Gore, further qualifying the consensus that is currently in favor of drastic global attenuation of human impacts.

The Darwin Mission, scheduled for launch in about 2015, is intended for planet detection, but will essentially drill down to search for atmospheres likely to support and to have been manufactured by life. James Lovelock was the first to suggest this potential for Mars,[7] proposing that life can maintain unbalanced atmospheres formed from mixtures of gaseous compounds that would otherwise be incompatible. It is perhaps understandable, then, that our best indicator of global ruination is an extreme imbalance in the already dynamic proportions of our planet's atmospheric gases. David King[8] has been watching these signs while raising concerns over climate change:

> [M]an's evolution into modern socioeconomic societies has created a raft of risks for the Earth's ability to provide the resources needed by those societies and by other life-forms. My view of the Earth system, as with Lovelock's *Gaia,*[9] is of a self-organizing system, far from equilibrium, with co-dependence and co-evolution of the geological, ocean, atmospheric, and biological systems. The characteristic of such a system is the potential for instability. [...] It doesn't take much to upset such a finely balanced system. Of course, our population spurt to the current 6 billion, coupled with our altering land use and fossil-fuel usage, has done just that. We are now entering a climate period which can be described as anthropocene, with severe potential consequences for our civilization over this millennium.
>
> But we are a species that has evolved a conscious ability to analyze our situation and to act on the analysis. [...] The big question for the well-being of future generations is whether or not our multifaceted, multicultural, [...] socio-political systems can rise to it—a challenge on an unprecedented scale.

This all seems straightforward enough. The governments of the world are telling us that we need to act fast to attenuate our environmental impacts. But a few are raising the possibility that the situation isn't quite as drastic as it first appeared, and others have suggested that our response to climate change might affect our future evolution in counterintuitive ways.[10] Such dissension from this verdict, shared by the majority, has attracted vitriol, as of course did *The Origin*. Bjørn Lomborg[11] once sparked a "firestorm debate" by suggesting that Earth's resources were more persistent than was suggested by the pessimistic forecasts of the preceding decade. His optimism extends to the new challenges posed by climate change:

> [M]aybe we're heading for doom. That is a possibility. But the choice that we have to make every day is, "Well, what can we do?" Are we going to be able to do a lot of good, or a little good? And so it doesn't really help to go, "Are we doomed?" [...] Bottom line is, you can't predict the future. You can only say, well we've got reasons

to believe that if the future looks anything like the past, which is the only thing we have to go by, then there's good reason to believe that we generally tend to solve more problems than we create. But that's not a totally satisfying answer and therefore we try to solve, or make models for individual circumstances. But we're just very, very poor at dealing with the fact that we do innovate.

Some would say that this lack of appreciation of human capacity to solve more problems than we create itself generates serious problems for society. If we cannot solve problems, then we are effectively slaves to nature's rhythms. Most would say that we are guilty of enslaving ourselves to the environment through our technological successes, at the expense of natural processes: impacts we call climate change, deforestation, pollution, etc. The popular response has been for considered mitigation of nature-impacting activities, reducing carbon footprints and emissions, recycling, and renewable energy. But Stuart Blackman[12] argues that such environmentally deterministic thinking, ironically yet inevitably, will lead to our history being determined by that very environment: "An unfounded sense of crisis dominates public discussion of environmental issues, and shrill demands for urgent action to mitigate climate change thrive at the expense of genuine, illuminating, nuanced debate about how to make the best of an uncertain future." He says:

> The consensus view that we mitigate against anthropogenic climate change has important implications for both future human history and our future evolution. The story of human history to date has been one of distancing ourselves from nature. After all, it's our civilization—our *development*—that has served increasingly to buffer us from the elements. And yet an emphasis on mitigation would likely serve to reverse that trend. Those who talk in terms of preventing climate change also tend to see development as the problem. The result of that way of thinking is that the human race would be left in a position where we are more vulnerable to whatever Mother Nature has to throw at us— and you can be sure she has plenty to throw, whether or not our industrial emissions are influencing the climate. In this way, environmentalism is a

self-fulfilling prophecy. By bringing us closer to nature, it exposes us to environmental dangers and potentially makes natural selection an important driving force once again.

This is quite a challenging viewpoint in contrast to all the publicity and politicking about global warming, but it's probably a healthy and necessary challenge; otherwise, we could blindly blunder on unchecked. But, regardless of the outcomes of this debate, in the longer term, there isn't much that human ingenuity and innovation can do to mitigate against natural catastrophes independent of mankind: meteor impacts, or our galaxy being engulfed by a roving black hole— our planet reduced to an infinitesimal dot, a mere morsel for an intergalactic Pac Man. Or, perhaps our end will be at the hand, tentacle, or sucker of another life-form, "but not as we know it, Jim"— Earth destroyed to make way for a hyperspace bypass.[13] It's impossible to predict, and they're not called crystal balls for nothing.

Humans, real and imagined, have often looked to the stars for answers to larger-than-life questions: from Ptolemy to Fred Hoyle, Moon-Watcher (*2001: A Space Odyssey,* by Arthur C. Clarke) to Dr. Eleanor Ann Arroway (*Contact,* by Carl Sagan). While we can be confident of our earthly explanations, outer space is where our logic breaks down, or when theoretical physics is forced into fantastical realms. Curvedness of the space-time continuum, the Poincaré homology sphere, and the Picard horn are all things that the 2008 inauguration of the Large Hadron Collider hopes to clarify. Mining the very fundament of the universe will reveal the basic relations of matter and how superstructures, including the universe, were formed. But, like looking at a pudding long after the cook is gone, understanding how the firmamental pavlova appeared is somewhat different from understanding *why* it got put there. If the Higgs boson (aka the "God Particle") is intergalactic gelatin, then who or what is beating the egg whites, whipping the cream, and ultimately scoffing it down in a fruit-laden feeding frenzy? Obviously, it has not escaped notice that turnover of inorganic, galactic matter is akin to an organic process, a natural process. A Darwinian process. And, as Darwinian evolutionary biology makes no comment on pangenesis, this new "evolutionary cosmology" does not attempt to explain the big question of why there is anything, rather than nothing. Notwithstanding such limits to our knowledge, could the universe really conform to the same evolutionary gradualism seen here on Earth? Even before his childhood's end, Richard L. Gregory[14] had perceived the existence of such parallels:

> I was brought up with evolution of the stars, as my father was an astronomer. As a boy in the 1930s, I would read Eddington and Jeans [the co-founders of British cosmology] with avidity; but although Darwin was quite often discussed, natural selection was at that time controversial and generally viewed with suspicion. The concept of design by random events, with successes and failures writing the future, was hardly appreciated, certainly not by me. Natural Selection is sometimes described as mindless and lacking intelligence—but it seems to me now that the Darwinian processes are intelligent, super-intelligent, producing answers science can hardly formulate, let alone fully understand.
>
> When Darwinian evolution is claimed to solve practically all problems of the universe, one has to ask: How did stars come into being? Darwin himself realized that his biological theory does not extend to the inorganic world, so regretfully [it] leaves problems of creation and development of lifeless matter an inscrutable mystery. Martin Rees tries to bridge this gap by thinking of something akin to organic evolution for the universe itself—successive creations and destructions gradually evolving the natural laws and matter. This is a wonderful idea. It remains to be seen whether Darwin's great insight for biology extends to the universe itself.

The idea of "successive creations and destructions gradually evolving the natural laws and matter" makes reference to Alan Guth's and Andrei Linde's ideas on a multiple universe, or multiverse, itself built upon the idea of a bubble universe.[15] Lee Smolin's evolutionary cosmology then posits an evolutionary mechanism underlying universe survival. Confused? You will be, but that's cosmology for you. To partially explain: If energy fluctuations in the parental "quantum foam," the vacuum precursor of universes, exceeds a certain threshold, then an expanding, persistent bubble universe forms. If not, then a small, temporary universe blips and dies in a single heartbeat of the eternal space-time continuum. Bubble universes like our own that do survive form matter and galactic structures, and can even propagate their own bubble children through the collapse of black holes. It follows that, the more black holes a universe contains, the more offspring it can spawn and the longer it will persist. So we have variation and a selective mechanism, the prerequirements for Darwinian evolution, and so, for Smolin's cosmological natural selection theory of fecund universes.

However, because each bubble arises from a fluctuating energy source, each descendant universe within the multiverse is likely to exhibit differing parameters, those physical constants and laws that we hear are so critical in their range of values to allow the existence of life: a fine-tuned universe. Statistically, there will be mostly bubbles with no life, but many fewer ought to have life like ours, and different from ours, perhaps so complex to be beyond imagination. A certain famous song[16] seems to sum it up rather well:

The universe itself keeps on expanding and expanding, in all of the directions it can whiz.

As fast as it can go,

that's the speed of light you know;

twelve million miles a minute, that's the fastest speed there is. So remember when you're feeling very small and insecure, how amazingly unlikely is your birth,

and pray that there's intelligent life somewhere up in space, 'cause there's bugger-all down here on Earth!

Let us hope that heightened intelligence, or at least common sense, is a part of human futures, perhaps so that humans can have any future prospects at all. Science fiction often predicts a dystopian future of tyranny and degradation, runaway technologies, clones, and postapocalyptic mutants. If anthropogenic calamities, like that forecast for global warming, can be averted and avoided, then perhaps science can provide a more optimistic outlook. Martin Rees[17] has great confidence in our potential, but not necessarily in our present, human form,

> Most educated people are aware that we're the outcome of nearly 4 billion years of Darwinian selection. But many tend to think of humans as somehow the culmination of this process. Astrophysics tells us, however, that our Sun is less than halfway through its life span. It will not be humans who watch the Sun's demise 6 billion years from now: Any creatures that exist then will be as different from us as we are from bacteria or amoebae. There's more time ahead, for future evolutionary change, than the entire emergence of our biosphere has needed. Moreover, evolution is now occurring not on the traditional timescale of natural selection, but at the far more rapid rate allowed by modern genetics, intelligently applied. And post-human life has abundant time to spread through the galaxy and beyond. Even if intelligent life is now unique to Earth, it could nonetheless become a significant feature of the cosmos. Our tiny planet could then be cosmically important as the "green shoot" that foliated into a living cosmos.

> I believe we are part of some marvelous evolutionary process which still has a long way to go beyond the human stage, here on Earth and far beyond. ... Extraterrestrial life will use genetic engineering to quickly modify themselves into new post-human species better adapted to an alien habitat.

Perhaps an altogether stranger alternative to colonizing alien habitats is the idea of a man-made universe, most easily conceived as a dreamlike computer simulation. If space exploration is all about externalizing our percepts, then whatever constitutes our consciousness sets about internalizing them. One aspect of this is our sense of self-awareness, which is altogether a result of feedback from our habitat. In order to fabricate that habitat, our senses must be duped with enough information that we are not to be left with any suspicions about its authenticity. This is somewhat aided by René Descartes' celebrated "cogito ergo sum" (French: *Je pense, donc je suis;* English: I am thinking, therefore I am) and its extension to his "truth rule," which states, "whatever I perceive very clearly and distinctly is true." Thus, the human tendency is to conflate experiential existence with our own existence; we trust our senses and commit our brains to the unconditional processing of their sensory harvest. So, even if we are unwilling subjects of an all-encompassing video game, we are likely to accept it as real, until a malfunction raises our suspicions that all is not as it seems.

Descartes' next step was to claim that, without knowing God exists, he could not be certain of any of his knowledge, because God is the source of his clarity. Unfortunately this sets up a circularity of reasoning, a Cartesian circle that can get very complicated, very quickly, as Peter Cook and Dudley Moore[18] once discovered with humorous consequences:

Dud: Are we in fact merely a reflection of ourselves as seen in a pool at twilight?

Pete: What you're saying is, if the imagination of an imagined being imagines that life itself is imaginary, how can the imagined life of the being who is himself imagined be imagined by the being who is imagining himself through a glass darkness. That's what you mean isn't it?

[*silence*]

Dud: Errm.

[*further silence*]

Dud: Yeah.

This all seems quite fantastic, in a *Through the Looking-Glass, and What Alice Found There*[19] *sort of way. And yet a strong case has been made for such an imaginary universe, possibly most useful as a check on our reality. This simulation hypothesis throws up many complexities for Darwinists. It throws up many complexities for anyone! Here are some immediate thoughts.*

First off, to implement this parallel state you would have to be hardwired into the system. Then what if your virtual personality was likewise hooked up to a game within the game? If you're an online gamer, then a quick bit of arithmetic is required here; basically, your Second Life avatar just got relegated to a "Third Life." Where this chain of virtual realities leads is mind-boggling, not least the possibility of recursion *ad nauseam*. But sticking with just one reality and one virtual reality, for beings within that simulation, evolution is as much a construct as everything else in the simulation, but that doesn't stop it feeling "natural" to us. For the simulation operators, natural selection is an algorithm that can be parameterized in order to introduce adaptations. For the scenario with beings plugged into the system, selection pressures could act to turn us into blobs to conserve energy and maximize brain function, depending on how our brains are wired and if the rest of our bodies are inactive. However, the main reason for adaptation in sexually reproductive organisms is to maximize the number of their offspring, but outside the system, there is no opportunity to reproduce and therefore no mechanism for evolution. This is all very confusing. We need Nick Bostrom[20] to explain his simulation hypothesis:

> If we are in a simulation, it is an open question what the world looks like outside the computer in which we are implemented. There would have to be an extremely advanced form of intelligence capable of creating such a simulation. Whether Darwinian evolution operates there would depend on whether its preconditions are in place.[21]

But you will probably be familiar with the basic idea of being wired into a grid, and being represented by an avatar within its internal world, from the blockbuster movie *The Matrix*. In this film, the lead character Neo (played by Keanu Reeves) is able to use the socket at the back of his head that connects him with "the Matrix" grid to upload knowledge directly into his brain: useful everyday skills, like Kung Fu and bullet-dodging. In Bostrom's silicon world ruled by software algorithms, similar artificial enhancements may be the only way to keep abreast of the competition:

> [Today] natural (and sexual) evolution of humans still occurs. However, because of our long generation span, it is imperceptibly slow and insignificant compared to other sources of change in the modern world. Our culture, economy, and technology now change significantly within a single generation, while significant biological change through evolution requires many generations.
>
> In the coming decades, genetic screening and modification will be developed and increasingly applied. If nothing else happened, then evolutionary change for humanity, 50 years from now, would be primarily driven by the application of these

new technological abilities. This would not mean that evolution would have ceased. If we imagine a world with advanced genetics persisting for a long time, evolution would eventually select for humans who used genetics to maximize their inclusive fitness. For example, the desire to have children has genetic correlates, and we would eventually evolve to have a stronger urge to have large numbers of offspring. (Currently, we mainly desire various proxies for children, such as love and sex, which in the past were highly correlated with the production of children.) Moreover, if genetic enhancements such as intelligence, health, or beauty correlate with a greater ability to acquire the resources needed to have many children, then we might also eventually evolve a propensity to make greater use of such genetic tech-nologies as are available to give our offspring these enhanced traits.

However, technological development does not end with advanced genetics. Before this century is up, it is likely that we will have developed mature nanotechnology as well as machine intelligence surpassing that of humans (perhaps by reverse-engineering the human brain). At this point, the driving force will no longer be biological human brains, but artificial intelligences or uploaded humans. These will develop much faster than biological humans, because they are not limited to hydrocarbon-based chemistry and our slow biological neural wetware.

Since artificial intelligences and uploads are software, their generation cycle can be extremely short: They can reproduce almost instantaneously, as quickly as one can make a copy of a computer program. In a population of human uploads, evolutionary selection would again kick in. Uploads who liked making copies of themselves would quickly proliferate, and, since their "children" would share the mental attribute of their progenitor, they too would like to make copies of themselves. This leads to very rapid exponential population growth, which in a matter of days could fill up all available computing space. Population growth would only plateau when the replicators run out of computing power. Selection would favor the most economically productive uploads or AIs, since they would acquire most of the resources needed to replicate or to accelerate or enhance their own performance. What the quality of life would be for productivity-optimized uploads is not known.

The only way to exercise long-term *control* over the direction of our evolutionary development is through global coordination. If only a subset of communities or countries decided to deviate from the fitness-maximizing path, evolution would simply move to a higher level, and these communities would eventually be out-competed by others. In the long run, therefore, it might be desirable to implement a global policy that could steer the evolution of intelligent life in a direction that maximizes its well-being. There is no guarantee that the default evolutionary outcome would be optimal or even acceptable from the point of view of realizing our human values.

Evolution is not normative. It is simply a factual constraint that may in the future again become directly relevant for humans or our successors, and that we could in principle work around through planning and global coordination.

This paper[22] has briefly looked at human relations with our earthly habitat as mediated by Darwinian evolution. Societies tend to reverse the effect of natural selection, releasing the pressures exerted on us by a dynamic environment. Ironically, our own impacts may be re-exposing us increasingly to those pressures. The advice is to modify our behaviour and be confident in our abilities to adapt to the threat, but there is also the danger of unforeseen consequences. Casting our gaze farther afield, deep into the evolving universe, we may yet discover alternative environments, but the Earth-bound scenarios for humanity in the future seem twofold: (1) adapting our behaviours that impact the environment, and (2) artificially adapting ourselves to environmental change. It seems that the take-home message from both looks to some sort of worldwide effort toward coordinating all our futures.

Notes

1. Charles Darwin, *On the Origin of Species by Means of Natural Selection, or the Preservation of Favoured Races in the Struggle for Life* (London: John Murray, 1859), p. 441.

2. The process by which characteristics that assist in surviving to reproductive maturity get passed to the next generation. The pressure is exerted by the environment to adapt to changing conditions.

3. "The Darwin Awards salute the improvement of the human genome *by honoring those who accidentally remove themselves from it." I personally find the ridicule of misfortunate victims of accidents distasteful, even if it is through their own stupidity. There is a criticality in the humor: the point at which someone is seriously hurt or worse having slipped on their banana skin, even if they put it there themselves—e.g, "A man is eating a banana. He throws away the skin and then slips on it, thus becoming the master of his own downfall in an ironic and amusing fashion." Not so funny if, as a consequence, his head is impaled on a railing. However, the humor in a tragic situation may be redeemed by a surreal, subsequent event: A popular example is when a man is mowing his lawn in open-toed sandals, chops his toe off, a terrible accident, even if stupid, until we are told that the toe shot up and took out his eye!*

4. Patrick Tort is director of the Institut Charles Darwin International (http://www.darwinisme.org) and professor at the Museum national d'Histoire naturelle in Paris. His publications include *Darwin et la Science de l'Évolution* (2000), *La Seconde Révolution Darwinienne (Biologie Évolutive et Théorie de la Civilisation)* (2002), *L'Effet Darwin: Sélection Naturelle et Naissance de la Civilisation* (2008), and *Bicentenaire de la Naissance de Charles Darwin* (2009, CD-ROM).

5. Charles Darwin, *The Descent of Man, and Selection in Relation to Sex* (London: John Murray, 1871).

6. Organisms that modify the natural selection pressures they encounter by their actions and choices. See F. J. Odling-Smee, K. N. Laland, and M. W. Feldman, *Niche Construction: The Neglected Process in Evolution*. Monographs in Population Biology. 37. Princeton University Press, 2003.

7. J. E. Lovelock, "A Physical Basis for Life Detection Experiment," in *Nature* 207 (1965), 568-570.

8. Sir David King, ScD, FRS, is the director of the Smith School of Enterprise and the Environment at the University of Oxford. He was the UK government's chief scientific adviser and head of the Government Office of Science from October 2000 to December 31, 2007, during which he raised the profile of the need for governments to act on climate change and was instrumental in creating the new £1 billion Energy Technologies Institute. He is co-author of *The Hot Topic: How to Tackle Global Warming and Still Keep the Lights On* (with Gabrielle Walker, 2008).

9. J. E. Lovelock, *Gaia: A New Look at Life on Earth* (Oxford University Press, 1979).

10. E.g., *Global Warming Could Be Reversing A Trend That Led To Bigger Human Brains* (University at Albany, Science Daily, March 23, 2007). Available online at http://www.sciencedaily.com /releases/2007/03/ 070322142633.htm.

11. Bjørn Lomborg is an adjunct professor at the Copenhagen Business School, director of the Copenhagen Consensus Centre, and a former director of the Environmental Assessment Institute in Copenhagen. He became internationally known for his best-selling and controversial book, *The Skeptical Environmentalist: Measuring the Real State of the World* (Cambridge University Press, 2001).

12. Stuart Blackman shares a "particular interest in the relationship between science and

politics" with Ben Pile, his co-editor of Climate Resistance: Challenging Climate Orthodoxy (http://www.climate-resistance.org/), which they base on the argument that "Environmentalism is in the ascendant. It holds that instead of buffering ourselves against whatever Mother Nature has to throw at us, we should try to make the weather marginally different by cutting down on the things that make life worth living."

13. A. Douglas Adams's *Hitchhiker's Guide to the Galaxy* reference: "People of Earth, your attention please. ... This is Prostetnic Vogon Jeltz of the Galactic Hyperspace Planning Council. ... As you will no doubt be aware, the plans for development of the outlying regions of the Galaxy require the building of a hyperspatial express route through your star system, and regrettably your planet is one of those scheduled for demolition. The process will take slightly less than two of your Earth minutes. Thank you."

14. Richard L. Gregory co-founded the Department of Machine Intelligence and Perception, a forerunner of the Department of Artificial Intelligence at the University of Edinburgh (with Donald Michie and Christopher Longuet-Higgins) in 1967, the same year that he presented the Royal Institution Christmas Lecture, *The Intelligent Eye*. He was made a Fellow of the Royal Society of Edinburgh in 1969 and a CBE in 1989, and he was elected to be a Fellow of the Royal Society in 1992, the same year that he was awarded the Royal Society Michael Faraday Medal, while he received the Hughling Jackson Gold Medal from the Royal Society of Medicine in 1999. He is currently an emeritus professor of neuropsychology at the University of Bristol. His contributions to TV and radio are extensive, as well as the design of science exhibitions, most notably, Hands-On Science at the Bristol "Exploratory." His books include *Mirrors in Mind* (1997) and *The Mind Makers* (1998).

15. For an overview see M. Tegmark, "Parallel Universes," in *Science and Ultimate Reality: From Quantum to Cosmos*, edited by J. D. Barrow, P. C. W. Davies, and C. L. Harper (Cambridge University Press, 2003). Available online at http://www.wintersteel.com/files/ShanaArticles/ multiverse.pdf.

16. The "Galaxy Song" by Eric Idle and John Du Prez, from Monty Python's *The Meaning of Life*.

17. Martin Rees delivered the 2007 Gifford Lecture, *21st Century Science: Cosmic Perspectives and Terrestrial Challenges*, at the University of St Andrews, and has taken part in the Edinburgh International Science Festival. His full form of address is Professor Sir Martin John Rees, Baron Rees of Ludlow, of Ludlow in the County of Shropshire. He has been president of the Royal Society since 2005 and is also Master of Trinity College, a professor of cosmology and astrophysics at the University of Cambridge, and visiting professor at Leicester University and Imperial College London. He was knighted in 1992, appointed Astronomer Royal in 1995, nominated to the House of Lords in 2005 as a cross-bench peer, and appointed a member of the Order of Merit in 2007. His current research deals with cosmology and astrophysics, especially gamma-ray bursts, galactic nuclei, black hole formation, and radiative processes (including gravitational waves), as well as cosmic structure formation, especially the early generation of stars and galaxies that formed at the end of the cosmic dark ages more than 12 billion years ago, relatively shortly after the Big Bang. His recent awards include the Royal Society's Michael Faraday Prize and lecture for science communication (2004), and the Royal Swedish Academy's Crafoord Prize (2005). Other notable awards include the Heinemann Prize (1984), the Balzan Prize (1989), the Bower Award of the Franklin Institute (1998), the Einstein Award from the World Cultural Council (2003), and the UNESCO Neils Bohr Medal (2005). He has authored or co-authored about five hundred research papers. He has lectured, broadcast, and written widely on science and policy, and is the author of seven books for a general readership, most recently, *What We Still Don't Know* (2007).

18. *Not Only ... But Also*, BBC, 1965.

19. C. L. Dodgson/L. Carroll, 1871.

20. Nick Bostrom is professor and director of the Future of Humanity Institute in the Faculty of Philosophy at Oxford University. He has previously held positions at Yale University and as a British Academy postdoctoral fellow. His doctoral dissertation, which was selected for inclusion in the Outstanding Dissertations series by the late Professor Robert Nozick, developed the first mathematically explicit theory of observation selection effects. He has published more than 140 articles in both philosophy and physics journals. He is the author of *Anthropic Bias: Observation Selection Effects in Science and Philosophy (2002)*.

21. N. Bostrom, "The Future of Human Evolution," in *Death and Anti-Death,* edited by C. Tandy (Ria University Press, 2004). Available online http://www.nickbostrom.com/.

22. All opinions are those of the author unless clearly indicated otherwise. Unless the literary source is otherwise indicated, all quotes were obtained through direct communication with each named contributor.

In which we counter a certain claim about the value of religion,

and end up on a road journey.

In an October article in *Aeon Magazine* (*Curb your enthusiasm* http://www.aeonmagazine.com/world-views/michael-ruse-humanism-religion), accompanied by a précis piece in *The Guardian*'s online *Comment is Free: Belief* (*Why Richard Dawkins' humanists remind me of a religion* http://www.guardian.co.uk/commentisfree/belief/2012/oct/02/richard-dawkins-humanists-religion-atheists), Michael Ruse argued for the coexistence of science and religion on the basis that they are, "asking different questions". I do not agree.

To accept that a religious mindset is appropriate in asking questions about the natural world at all, you are forced to also accept its trappings of revealed wisdom and an untestable premise that there exists an unerring, higher source of knowledge. Without these, religion ceases to be religion. What is left, in its rawest, unadulterated form, is called "philosophy".

Religion is clothing; a veil. Draped over philosophy, it became theology. Theology draped over geology, zoology and botany became natural theology. That is how christianity developed it's position in opposition to natural science, by starting from a religious source, emanating from an entirely different school to that of scientific naturalism.

It is no mystery that science and religion have different starting points. For example, the UK's Chief Rabbi, Lord Sacks, who believes that science, "is one of the greatest achievements of humankind, a gift given to us by God" (BBC 2012 *Rosh Hashanah: Science vs Religion*), defines it as, "science takes things apart, to see how they work", whilst, "religion puts things together, to see what they mean". From this dichotomy arises the single most used apology (*sensu* "argument") for religion and the one that Ruse forwards, that religion can answer questions that science cannot.

This preconception is an important but often ignored conditioning of people. Their psychological predispositions are the difference that initially established and now perpetuate a polarised science-religion debate. Reasoned arguments are ineffectual because, in the minds of different people, the architecture of reason is constructed differently. Both cannot, by definition, be correct. It is quite clear from bountiful evidence that scientific logic is not only better founded, but will also provide a portal into dissecting irrational psychologies: the science of religion. The obverse cannot be claimed.

A better understanding of the psychology of faith is needed, because the divide will persist until there is sufficient comprehension of the irrational mindset. For this we must again rely upon scientific disassembly. On this point is where Ruse's article and I do agree; in that same article he quotes from renowned naturalist, E.O. Wilson's *On Human Nature*,

> *... we have come to the crucial stage in the history of biology when religion itself is subject to the explanations of the natural sciences. As I have tried to show, sociology can account for the very origin of mythology by the principle of natural selection acting on the genetically evolving material structure of the human brain.*
>
> *If this interpretation is correct, the final decisive edge enjoyed by scientific naturalism will come from its capacity to explain traditional religion, its chief competition, as a wholly material phenomenon.*

Human brain evolution has produced an inherently curious and intelligent reasoning that from a very early age seeks to identify causes for effects. Coupled with a propensity for patterns and design, it is hardwired into our cognitive faculties to recognise structural design in nature and interpret a purpose for natural phenomena. While reason and logic are adaptions towards problem solving, myths and religions are likely secondary products of our interpretation of the natural world, sometimes known as "spandrels" (*sensu* Gould & Lewontin: a byproduct of an adaption).

Over the course of human history, civilisations have used metaphors as indirect descriptors of their direct environment, to gloss over where information has been lacking. Prior to the Enlightenment, this has resulted in a host of supernatural explanations for natural dynamics, each associating change with purpose. People continue to be comforted in these explanations and communities strengthened, encouraged that everything happens for a reason and not at the whim of chance. It is no coincidence therefore that there is a close tally between the number of these stories and the number of cultural contexts that bore them.

Analogous to Darwin's "Tree of Life", the multitudinous faiths have evolved over time, diversifying, bifurcating and running in parallel, as related yet independent branches of thought. Some have died out and others have persisted. Some are closely related while others are quite alien, but what is consistent is how parochial all these stories were to their original context, how they relied upon reference to their nearby surroundings to construct their narrative, how they only attempted to make sense of their immediate environment. Psychologist and theologian, Justin L. Barrett expresses it in terms of,

> *the way our minds solve problems generates a god-shaped conceptual space waiting to be filled by the details of the culture into which we are born.*

Despite the obvious origins of faiths, and as a footnote to sceptics and atheists, there is a Catch-22 hypocrisy in accepting the scientific evidence that myths and religions have emerged over time as products of our evolutionary history. It's the having your cake and eating it type: as religion is an evolutionary byproduct, then it's only nature and reacting against it is, well, in one sense of the word, unnatural.

While a large number of people do apply logic and conclude that religion doesn't add anything to their world for whatever personal reasons, (usually because it fails to offer any further explanations beyond the laws of biology, chemistry and physics that comprehensively describe our universe), in contrast, most people in the world are religious, following one or other religion or faith system.

This may be mystifying, but it is not surprising. People are genetically programmed to do so. Indeed, it would be more surprising if they were not religious and the reasons given for adhering to religions reflect this innate urge: "I feel it in my bones", "it comes from within". While there is no evidence towards proving the existence of deities, there are however many cultural and social mechanisms that act throughout a lifetime to reinforce religion. In response, societal goals of a more rational understanding of the universe and a more rational approach to life, need a better education towards a greater science literacy.

Being scientifically informed doesn't mean that we can understand every aspect of the most technical problem, from particle physics to plate tectonics, but instead provides us with a basic toolkit of knowledge and skills about science and technology, and a way to look at the world. We can use the tools in this kit to better inform everything from living our daily lives to running whole nations.

Any good tool should provide a metric, some measure by which comparisons may be made. The one I feel most ideal is at the heart of the science-religion debate: the dissent over our origins as a dissociation from nature. Thus, to investigate human tendencies towards faith, think of a continuous road that stretches beyond the horizon in both directions. Close at hand we shall place a recognisable marker to identify a point by which all others can be measured in scale. The scale is defined through an ability to explain natural phenomena. Namely, nature in all it's manifestations.

On the Origin of Species is essentially about the generation of biodiversity in nature. Darwinian evolution also features predominantly in arguments refuting the biblical accounts of human origins. However, Darwinism has also been stretched far beyond its original scope. It has been applied to a cornucopia of human behaviour, from entrepreneurialism to war crimes, and held up as an answer or a scapegoat in countless situations. To differing degrees, it is an important part of how we understand ourselves, our history and our culture. So, let us place Mr Darwin here as a totem for neo-Darwinism, the most comprehensive acceptance of his ideas, and rank the range of interpretations of Darwinism in order of their loyalty to those original ideas.

To achieve this I've interviewed over fifty commentators: conservationists and creationists, bishops and biochemists, palaeoceanographers and Intelligent Design theorists, theistic evolutionists and a Bahá'í lecturer, sex researchers, mathematicians, ophthalmologists, linguists, evangelical Christians, philosophers, physicians and the Astronomer Royal. As a starting point, I asked each one the same question: 'what does Darwin mean to you as an individual, and as part of humanity?'

Contributors to date include, Richard Dawkins, Noam Chomsky, Oliver Sacks, James Watson, Ian Stewart, Edward Wilson, Martin Rees, Simon Conway Morris, David King, Aubrey Manning, Michael Behe, George Schaller, Brian Charlesworth, Bjørn Lomborg, Daniel Dennett, William Dembski, Stephen Wolfram, Rupert Sheldrake, Michael Ruse, Susan Blackmore, Lewis Wolpert, Steven Pinker, Richard Holloway, Richard Lewontin, Randal Keynes, John Polkinghorne, Tim Smit, Matt Ridley, Archimedes Plutonium, Richard Gregory, Ken Ham, Adrian Hawkes, and many more. Furthermore, I hope soon to add our host Robin Ince to that august gathering.

Now, on moving away from Darwin we can position each alternative interpretation along our road, measured out in units of "distances from Darwin". Our journey will take us quite some distance away from Darwin until he is but a dot on the horizon from which we set out. But the order in which we distribute alternative interpretations *en route* may not accord to a preconceived sequence reinforced by the polemics that usually dominate this debate. Much of what I've discovered is surprising and exciting - preconceptions are challenged, antagonists are revealed to be uncomfortable bedfellows, and the extremists aren't necessarily who you might think they are.

A them-and-us approach has resulted in a stand-off. Clearly there is more complexity involved in the science-religion debate. When we are done with our gradient, we will be able to look back along our journey and recognise the true diversity of understanding that really describes a continuum spelled out by individuals standing shoulder-to-shoulder along our road. Nonetheless, we will have our path running from one extreme to the other and between them a Distance-from-Darwin gradient that traverses the rich and fertile landscape of human thought.

Dissecting the Dissent of Man

Current understanding suggests that religion is partly an adaptive advantage by fostering cooperation between individuals. However, antagonistic to this strengthening of community relations, psychological predisposition for faith initially established and now perpetuates a polarised science-religion debate. Yet this evolutionary pre-programming of people is an important but often omitted element. It is also why reasoned arguments are ineffectual because, in the minds of different people, the architecture of reason is constructed differently. Both cannot, by definition, be correct. Hence the conflict.

Human brain evolution has produced an inherently curious and intelligent reasoning that from a very early age seeks to identify causes for effects. Coupled with a propensity for patterns and design, it is hardwired into our cognitive faculties to recognise structural design in nature and interpret a purpose for natural phenomena. While reason and logic are evolutionary adaptions towards problem solving, faith-based myths and religions are likely secondary products of our interpretation of the natural world. Consequently, as faith in the individual is an evolutionary by-product, there is a certain amount of hypocrisy in setting evolution against religion in order to criticise personal religious philosophies. Put another way, if faith is only nature in action, then reacting against it is, in one sense of the word, unnatural.

Why some people are religious and others are not throws up interesting questions about individual differences and histories, for which a burgeoning science of religion hopes to provide answers and a portal into dissecting alternative psychologies. In contrast to a superstitious outlook, being scientifically informed doesn't mean that we can understand every aspect of the most technical problem, from particle physics to plate tectonics. Instead, it provides us with a basic toolkit of knowledge and skills about science and technology, and a way to look at the world. We can use the tools in this kit to better inform human activity, from living our daily lives to running whole nations, and the case of religion is no different.

Any good tool should provide a metric, some measure by which comparisons may be made. The one most suited for our attempt to investigate human tendencies towards faith is already at the heart of the science-religion debate: the dissent over our origins as a dissociation from nature. Therefore, think of a continuous road that stretches beyond the horizon in both directions. Close at hand we shall place a recognisable marker to identify a point by which all others can be measured in scale. The scale is defined through an ability to explain natural phenomena, particularly human origins. Enter Charles Darwin and natural selection.

On the Origin of Species is essentially about the generation of biodiversity in nature. Darwinian evolution also features predominantly in arguments refuting biblical accounts of human origins. However, Darwinism has also been stretched far beyond its original scope. It has been applied to a cornucopia of human behaviour, from entrepreneurialism to war crimes, and held up as an answer or a scapegoat in countless situations. To differing degrees, it is an important part of how we understand ourselves, our history and our culture. So, let us place Mr Darwin here as a totem for neo-Darwinism, the most comprehensive acceptance of his ideas, and rank the range of interpretations of Darwinism in order of their loyalty to those original ideas. On

moving away from Darwin we can position each alternative interpretation along our road, measured out in units of "distance-from-Darwin". Our journey will take us quite some distance away from Darwin until he is but a dot on the horizon from which we set out, but hopefully it will reveal insights into human thought, some pertaining to faith.

Understandably, most scientists and the majority of biologists will be placed at a short distance-from-Darwin. Richard Dawkins is a well known Darwinist,

> Darwin is so important, it is almost absurd that children don't learn about it when they are tiny, practically, because it is the explanation for our existence, the existence of all living things, and it's in a way, one of the most powerful explanations of anything that anybody has ever suggested, because if you think about it, the ratio of that which it explains, which is everything about life and everything about complexity, divided by that which you need to postulate in order to do the explaining. It's a colossal ratio, because what you have to postulate is extraordinarily simple. It actually amounts to little more than high fidelity genetics because everything else kind of follows naturally. And from that almost miniscule level of assumptions you can explain just about everything about life, and yet, simple as the explanation is, powerful as it is, and huge as the magnitude of what it explains is, nobody thought of it before the middle of the 19th century. Which is an astonishing fact because it doesn't require great mathematics, doesn't require highly sophisticated notation of any sort, anybody can understand, although an amazing number of people fail to. So, it's almost unique, perhaps it is unique, in it's in the sheer power of what it does for the human intellect.

While here is another,

> Most of what Dawkins uses in regard to Darwin, of course we agree with. Because we agree with natural selection, we agree with speciation, we agree that mutations occur and so on ... Dawkins and us are very similar in very many ways.

Who do you think said this? Another biologist? Another scientist? In actuality, it was a young-earth creationist, Ken Ham the president of *Answers in Genesis* and the

Creation Museum in Petersburg, Kentucky.

Attitudes like this contradict the standard perception of the science-religion debate, that neo-Darwinist atheism forms one side and creationism the opposite, of which young-earth creationism is generally thought of being an extreme. Clearly a better understanding of the psychology of faith is needed in order to identify the underlying causes of dissension.

So, if there is significant agreement, then from where does the debated difference stem? Unsurprisingly, for young-earth creationists the problem lies with a 4.54 ± 0.05 billion year old planet and a Darwinian gradualist view of nature,

> for me Darwinian evolution is not really this issue so much as the millions of years. The age of the earth is the crucial issue ... Darwin needs time.

But here Ham reveals more than the root cause of dissension. He also indirectly identifies an important difference and perhaps the most fundamentally divisive influence that maintains the science-religion debate. Simply put, people differ in their starting points.

It is no mystery that science and religion have different starting points. For example, the UK's Chief Rabbi, Lord Sacks, who believes that science, "is one of the greatest achievements of humankind, a gift given to us by God" (BBC 2012 *Rosh Hashanah: Science vs Religion*), defines it as, "science takes things apart, to see how they work", whilst, "religion puts things together, to see what they mean". From this dichotomy arises the single most used apology (sensu "argument") for religion, that it can answer questions that science cannot. The issue for science isn't so much reductionism versus non-reductionist methods, but that creationism uses a starting point with an *a priori* assumption that a deity exists, while creationists like Ham claim science's starting point is that a deity does not exist,

> I think Dawkins is not prepared to accept that he has a starting point. He says that he works from evidence, and you're not allowed to work from the Bible. And if I was in discussion with him, I would have to force him to admit that he has a particular starting point. If he's not prepared to admit it, then I'd say he's inconsistent. I'd say that he's also inconsistent in that, if he's going to talk rationally and logically, we all accept the laws of logic and natural law, and the uniformity of nature, and so on, and I'd say it's illogical to do so if you don't believe in God.

The additional problem involves the differing standards for acceptance of the underlying evidence that informs those starting points. Creationists claim the Bible is evidence, but science dismisses it as inadmissible. In their turn, Darwinists forward *On the Origin of Species* as a flagship for Darwin's empirical work, and the wealth of data that have been collected since, but creationists question the scientific methods of measurement and interpretations of these findings. Thus, the overlaps between Darwinism and creationism appear in the details of shared understanding about evolutionary processes, but this is countered by their lack of sharing a common starting point for their world views. Evidence is at the heart of the matter, but what constitutes viable evidence is questioned by both sides of the debate.

In similar discord with Darwinists is a relatively recent derivative of creationism, the intelligent design movement. Intelligent design also accepts some aspects of evolution by natural selection, however, because of its outright rejection of Darwinian processes in the formation of some structures, which they claim are irreducibly complex,

William Dembski and his colleagues remove themselves even further from Darwin,

> Before Darwin, the power of choice was confined to designing intelligences - to conscious agents that could reflect deliberatively on the possible consequences of their choices. Darwin's claim to fame was to argue that natural forces, lacking any purposiveness or prevision of future possibilities, likewise have the power to choose. Accordingly, Darwin invented an oxymoron: natural selection. In ascribing the power to choose to natural selection, Darwin perpetrated the greatest intellectual swindle in the history of ideas. Nature has no power to choose. All natural selection does is narrow the variability of incidental organismal change by weeding out the less fit. Moreover, it acts on the spur of the moment, based solely on what the environment at the present time deems fit, and thus without any prevision of future possibilities. And yet this blind process, when coupled with another blind process, namely, incidental organismal change (which neo-Darwinians understand as genetic mutations), is supposed to produce designs that exceed the capacities of any designers in our experience ... Getting design without a designer is a good trick indeed. But with advances in technology as well as in the information and life sciences (especially molecular biology), it's a trick that can no longer be maintained. It's time to lay aside the smokescreens and the handwaving, the just-so stories and the stonewalling, the bluster and bluffs, and explain scientifically what people have known right along, namely, why the appearance of design in biology is not merely an appearance but in fact the result of an actual intelligence. This is the fundamental claim of intelligent design.

Furthermore, Darwinists and creationists are equally disparaging of intelligent design, accusing the movement of not carrying its arguments to the obvious conclusion and naming the source of intelligence. It is obvious that theirs is a teleological argument about which Ham says, "the intelligent design movement ... They go on about saying there has to be an intelligence behind life, and we would agree with that, but as creationists we're saying you cant stop there. We want to tell them who the intelligence is, that's the guy in the Bible."

Other faiths also adopt a variety of Darwinism, each adapted to their preconditions. Michael Cremo is a Hindu creationist, a religion populated by millions of avatars of their godhead Brahman,

> My Vedic alternative to Darwin's theory is not, however, in all ways the same as the theory of special creation of each species that Darwin argued against, nor is it in all ways different from Darwin's theory. Like Darwin's theory, my Vedic alternative explanation for the origin of species involves common descent from an original form (although that original form turns out to be Brahma, not a single celled organism). My Vedic alternative also involves a process of descent with modification, although not exactly the kind that Darwin envisioned.

Here a Hindu has applied Darwinism outside a purely biological context. Notably, Dembski previously accused biology of employing "just-so stories" in his attack on Darwinism, but interestingly, it has also been used in defence of Darwinism and criticism of other applications outside the original context. Richard Lewontin is a staunch critic of gene-oriented reductionism in the social sciences and other disciplines such as, evolutionary psychology, evolutionary ethics, evolutionary computation, evolutionary cosmology, memetics, digital Darwinism, corporate Darwinism and social Darwinism, *etc.*,

> The problem is that many have turned Darwin's description of the way in which organic evolution works into (1) a speculative tool for inventing a natural selective explanation for everything in the world, with no conceivable way of checking on the reality of these "Just So" stories and (2) have extended Darwin's structure which was tied to a particular natural phenomenology- the biological reproduction of offspring and the differential probability of survivorship and reproduction of those offspring in a real world of biological objects into a generalized metaphor for every kind of historical change in human culture and human history. This has led to the production of a vast literature on sociobiology, "evolutionary" psychology, "evolutionary" accounts of history, "evolution" of culture which are all intellectual games that vulgarize the Darwinian explanatory structure in the interest of producing a general theory of everything.

Evolutionary psychology has perhaps proven the most contentious field to receive the Darwin treatment. Arguments have revolved around such topics as units of selection, adaptation, hypothesis testing, extrapolation and political versus scientific purpose. These concerns would place it at a further distance-from-Darwin than that envisaged by Steven Pinker, one of its greatest advocates,

> Evolutionary psychology [is] the organizing framework—the source of 'explanatory adequacy' or a 'theory of the computation'—that the science of psychology had been missing. Like vision and language, our emotions and cognitive faculties are complex, useful, and nonrandomly organized, which means that they must be a product of the only physical process capable of generating complex, useful, nonrandom organization, namely natural selection ... Evolutionary psychology is changing the face of theories, making them into better depictions of the real people we encounter in our lives and making the science more consonant with common sense and the wisdom of the ages. Before the advent of evolutionary thinking in psychology, theories of memory and reasoning typically didn't distinguish thoughts about people from thoughts about rocks or houses. Theories of emotion didn't distinguish fear from anger, jealousy, or love. And theories of social relations didn't distinguish among the way people treat family, friends, lovers, enemies, and strangers.

A them-and-us approach has resulted in a stand-off, and clearly there is more complexity involved in the science-religion debate. A comparative gradient that acknowledges diversity in thought, rather than a parsimonious 2-sided debate, is informative in checking our assumptions and drawing out differences and similarities. When we are done with our gradient, we will be able to look back along our journey and recognise the true diversity of understanding that really describes a continuum spelled out by individuals standing shoulder to shoulder along our road. Nonetheless, we will have our path running from one extreme to the other and between them a distance-from-Darwin gradient that traverses the rich and fertile landscape of human thought.

www.ingramcontent.com/pod-product-compliance
Lightning Source LLC
Chambersburg PA
CBHW081046170526
45158CB00006B/1876